EVERYTHING BAD IS GOOD FOR YOU

EVERYTHING BAD
IS GOOD FOR YOU

*   *   *

How Today's Popular Culture
Is Actually Making Us Smarter

STEVEN JOHNSON

RIVERHEAD BOOKS
NEW YORK
2005

306.0973
J

RIVERHEAD BOOKS
Published by the Penguin Group
Penguin Group (USA) Inc., 375 Hudson Street, New York, New York 10014,
USA · Penguin Group (Canada), 10 Alcorn Avenue, Toronto, Ontario M4V 3B2,
Canada (a division of Pearson Penguin Canada Inc.) · Penguin Books Ltd,
80 Strand, London WC2R 0RL, England · Penguin Ireland, 25 St Stephen's
Green, Dublin 2, Ireland (a division of Penguin Books Ltd) · Penguin Group
(Australia), 250 Camberwell Road, Camberwell, Victoria 3124, Australia
(a division of Pearson Australia Group Pty Ltd) · Penguin Books India Pvt Ltd,
11 Community Centre, Panchsheel Park, New Delhi–110 017, India · Penguin
Group (NZ), Cnr Airborne and Rosedale Roads, Albany, Auckland 1310,
New Zealand (a division of Pearson New Zealand Ltd) · Penguin Books
(South Africa) (Pty) Ltd, 24 Sturdee Avenue, Rosebank,
Johannesburg 2196, South Africa

Penguin Books Ltd, Registered Offices:
80 Strand, London WC2R 0RL, England

Library of Congress Cataloging-in-Publication Data

Johnson, Steven, date.
Everything bad is good for you : how today's popular culture is
actually making us smarter / Steven Johnson.
p.   cm.
ISBN 1-57322-307-7
1. Popular culture.   2. Intellect.   I. Title.
HM621.J64   2005   2005042769
306'.0973—dc22

Printed in the United States of America
1   3   5   7   9   10   8   6   4   2
This book is printed on acid-free paper. ∞

Book design by Lovedog Studio

While the author has made every effort to provide accurate telephone numbers
and Internet addresses at the time of publication, neither the publisher nor the
author assumes any responsibility for errors, or for changes that occur after
publication. Further, the publisher does not have any control over and does not
assume any responsibility for author or third-party websites or their content.

For Lydia, true believer

Contents

SCIENTIST A: Has he asked for anything special?

SCIENTIST B: Yes, why, for breakfast . . . he requested something called "wheat germ, organic honey, and tiger's milk."

SCIENTIST A: Oh, yes. Those were the charmed substances that some years ago were felt to contain life-preserving properties.

SCIENTIST B: You mean there was no deep fat? No steak or cream pies or . . . hot fudge?

SCIENTIST A: Those were thought to be unhealthy. . . .

—from Woody Allen's *Sleeper*

Ours is an age besotted with graphic entertainments. And in an increasingly infantilized society, whose moral philosophy is reducible to a celebration of "choice," adults are decreasingly distinguishable from children in their absorption in entertainments and the kinds of entertainments they are absorbed in— video games, computer games, hand-held games, movies on their computers and so on. This is progress: more sophisticated delivery of stupidity.

—George Will

This book is an old-fashioned work of persuasion that ultimately aims to convince you of one thing: that popular culture has, on average, grown more complex and intellectually challenging over the past thirty years. Where most commentators assume a race to the bottom and a dumbing down—"an increasingly infantilized society," in George Will's words—I see a progressive story: mass culture growing more sophisticated, demanding more cognitive engagement with each passing year. Think of it as a kind of positive brainwashing: the popular media steadily, but almost imperceptibly, making our minds sharper, as we soak in entertainment usually dismissed as so much lowbrow fluff. I call this upward trend the Sleeper

Curve, after the classic sequence from Woody Allen's mock sci-fi film, where a team of scientists from 2173 are astounded that twentieth-century society failed to grasp the nutritional merits of cream pies and hot fudge.

I hope for many of you the argument here will resonate with a feeling you've had in the past, even if you may have suppressed it at the time—a feeling that the popular culture isn't locked in a spiral dive of deteriorating standards. Next time you hear someone complaining about violent TV mobsters, or accidental onscreen nudity, or the inanity of reality programming, or the dull stares of the Nintendo addicts, you should think of the Sleeper Curve rising steadily beneath all that superficial chaos. The sky is not falling. In many ways, the weather has never been better. It just takes a new kind of barometer to tell the difference.

Introduction

THE SLEEPER CURVE

EVERY CHILDHOOD HAS its talismans, the sacred objects that look innocuous enough to the outside world, but that trigger an onslaught of vivid memories when the grown child confronts them. For me, it's a sheaf of xeroxed numbers that my father brought home from his law firm when I was nine. These pages didn't seem, at first glance, like the sort of thing that would send a grade-schooler into rapture. From a distance you might have guessed that they were payroll reports, until you got close enough to notice that the names were familiar ones, even famous: Catfish Hunter, Pete Rose, Vida Blue. Baseball names, stranded in a sea of random numbers.

Those pages my dad brought home were part of a game,

though it was a game unlike any I had ever played. It was a baseball simulation called APBA, short for American Professional Baseball Association. APBA was a game of dice and data. A company in Lancaster, Pennsylvania, had analyzed the preceding season's statistics and created a collection of cards, one for each player who had played more than a dozen games that year. The cards contained a cryptic grid of digits that captured numerically each player's aptitudes on the baseball diamond: the sluggers and the strikeout prone, the control artists and the speed demons. In the simplest sense, APBA was a way of playing baseball with cards, or at least pretending to be a baseball *manager*: you'd pick out a lineup, decide on your starting pitchers, choose when to bunt and when to steal.

APBA sounds entertaining enough at that level of generality—what kid wouldn't want to manage a sports team?—but actually playing the game was a more complicated affair. On the simplest level, the game followed this basic sequence: you picked your players, decided on a strategy, rolled a few dice, and then consulted a "lookup chart" to figure out what happened—a strikeout, or a home run, a grounder to third.

But it was never quite that simple with APBA. You could play against a human opponent, or manage both teams yourself, and the decisions made for the opposing team transformed the variables in subtle but crucial ways. At the beginning of each game—and anytime you made a substitution—you had to add up all the fielding ratings for each player in your lineup. Certain performance results would change if your team was unusually adept with the glove, while teams that were less talented defensively would generate more errors. There were completely different charts depending on the number of runners on base: if you had a man on third, you consulted the "Runner on Third" chart. Certain performance numbers came with different results, depending on the quality of the pitcher: if you were facing a "grade A" pitcher, according to the data on his card, you'd get a strikeout, while a "grade C" pitcher would generate a single to right field. And that was just scratching the surface of the game's complexity. Here's the full entry for "Pitching" on the main "Bases Empty" chart:

> The hitting numbers under which lines appear may be altered according to the grade of the pitcher against whom the team is batting. Always observe the grade of the pitcher and look for possible changes of those numbers which are underlined. "No Change" always refers back to the D, or left, column and always means a base hit. Against Grade D pitchers there is never any change—the left hand column only is used. When a pitcher is withdrawn from the game make a note of the grade of the pitcher who relieves him. If his grade is different, a different column must be referred to when the underlined numbers come up. Certain players may have the numbers

7, 8, and/or 11 in the second columns of their cards. When
any of these numbers is found in the second column of a
player card, it is not subject to normal grade changes. Always use the left (Grade D) column in these cases, no
matter what the pitcher's grade is. Occasionally, pitchers
may have A & C or A & B ratings. Always consider these
pitchers as Grade A pitchers unless the A column happens
to be a base hit. Then use the C or B column, as the case
may be, for the final play result.

Got that? They might as well be the tax form instructions you'd happily pay an accountant to decipher. Reading
these words now, I have to slow myself down just to follow
the syntax, but my ten-year-old self had so thoroughly internalized this arcana that I played hundreds of APBA
games without having to consult the fine print. *An 11 in the
second column on the batter's card? Obviously,* obviously
*that means ignore the normal grade changes for the pitcher.
It'd be crazy not to!*

The creators of APBA devised such an elaborate system
for understandable reasons: they were pushing the limits of
the dice-and-cards genre to accommodate the statistical
complexity of baseball. This mathematical intricacy was
not limited to baseball simulations, of course. Comparable
games existed for most popular sports: basketball sims that
let you call a zone defense or toss a last-minute three-point
Hail Mary before the clock ran out; boxing games that let

you replay Ali/Foreman without the rope-a-dope strategy. British football fans played games like Soccerboss and Wembley that let you manage entire franchises, trading players and maintaining the financial health of the virtual organization. A host of dice-based military simulations re-created historical battles or entire world wars with painstaking fidelity.

Perhaps most famously, players of Dungeons & Dragons and its many imitators built elaborate fantasy narratives—all by rolling twenty-sided dice and consulting bewildering charts that accounted for a staggering number of variables. The three primary manuals for playing the game were more than five hundred pages long, with hundreds of lookup charts that players consulted as though they were reading from scripture. (By comparison, consulting the APBA charts was like reading the back of a cereal box.) Here's the *Player's Handbook* describing the process by which a sample character is created:

> Monte wants to create a new character. He rolls four six-sided dice (4d6) and gets 5, 4, 4, and 1. Ignoring the lowest die, he records the result on scratch paper, 13. He does this five more times and gets these six scores: 13, 10, 15, 12, 8, and 14. Monte decides to play a strong, tough Dwarven fighter. Now he assigns his rolls to abilities. Strength gets the highest score, 15. His character has a +2 Strength bonus that will serve him well in combat. Con-

stitution gets the next highest score, 14. The Dwarf's +2 Constitution racial ability adjustment [see Table 2-1: Racial Ability Adjustments, pg. 12] improves his Constitution score to 16, for a +3 bonus. . . . Monte has two bonus-range scores left (13 and 12) plus an average score (10). Dexterity gets the 13 (+1 bonus).

And that's merely defining the basic faculties for a character. Once you released your Dwarven fighter into the world, the calculations involved in determining the effects of his actions—attacking a specific creature with a specific weapon under specific circumstances with a specific squad of comrades fighting alongside you—would leave most kids weeping if you put the same charts on a math quiz.

Which gets to the ultimate question of why a ten-year-old found any of this *fun*. For me, the embarrassing truth of the matter is that I did ultimately grow frustrated with my baseball simulation, but not for the reasons you might expect. It wasn't that arcane language wore me down, or that I grew tired of switching columns on the Bases Empty chart, or that I decided that six hours was too long to spend alone in my room on a Saturday afternoon in July.

No, I moved on from APBA because it wasn't realistic enough.

My list of complaints grew as my experience with APBA deepened. Playing hundreds of simulated games revealed the blind spots and strange skews of the simulation. APBA neglected the importance of whether your players were left-handed or right-handed, crucial to the strategy of baseball. The fielding talents of individual players were largely ignored. The vital decision to throw different kinds of pitches—sliders and curveballs and sinkers—was entirely absent. The game took no notice of *where* the games were being played: you couldn't simulate the vulnerable left-field fence in Fenway Park, so tempting to right-handed hitters, or the swirling winds of San Francisco's old Candlestick Park. And while APBA included historic teams, there was no way to factor in historical changes in the game when playing teams from different eras against each other.

And so over the next three years, I embarked on a long journey through the surprisingly populated world of dice-baseball simulations, ordering them from ads printed in the back of the *Sporting News* and Street and Smith's annual baseball guide. I dabbled with Strat-o-Matic, the most popular of the baseball sims; I sampled Statis Pro Baseball from Avalon Hill, maker of the then-popular Diplomacy board game; I toyed with one title called Time Travel baseball that specialized in drafting fantasy teams from a pool of historic players. I lost several months to a game called Extra Innings that bypassed cards and boards altogether; it didn't even come packaged in a box—just an oversized envelope stuffed with pages and pages of data. You rolled six separate dice to complete a play, sometimes consulting five or six separate pages to determine what had happened.

Eventually, like some kind of crazed addict searching for an ever-purer high, I found myself designing my own simulations, building entire games from scratch. I borrowed a twenty-sided die from my Dungeons & Dragons set—the math was far easier to do with twenty sides than it was with six. I scrawled out my play charts on yellow legal pads, and translated the last season's statistics into my own home-brewed player cards. For some people, I suppose, thinking of youthful baseball games conjures up the smell of leather gloves and fresh-cut grass. For me, what comes to mind is the statistical purity of the twenty-sided die.

This story, I freely admit, used to have a self-congratulatory moral to it. As a grownup, I would tell new friends about my fifth-grade days building elaborate simulations in my room, and on the surface I'd make a joke about how uncool I was back then, huddled alone with my twenty-sided dice while the other kids roamed outside playing capture the flag or, God forbid, *real* baseball. But the latent message of my story was clear: I was some kind of statistical prodigy, building simulated worlds out of legal pads and probability charts.

But I no longer think that my experience was all that unusual. I suspect millions of people from my generation probably have comparable stories to tell: if not of sports simulations then of Dungeons & Dragons, or the geopolitical strategy of games like Diplomacy, a kind of chess superimposed onto actual history. More important, in the quarter century that has passed since I first began exploring those xeroxed APBA pages, what once felt like a maverick obsession has become a thoroughly mainstream pursuit.

This book is, ultimately, the story of how the kind of thinking that I was doing on my bedroom floor became an everyday component of mass entertainment. It's the story of how systems analysis, probability theory, pattern recognition, and—amazingly enough—old-fashioned *patience* became indispensable tools for anyone trying to make sense of modern pop culture. Because the truth is my solitary obsession with modeling complex simulations is now ordinary behavior for most consumers of digital age entertainment. This kind of education is not happening in classrooms or museums; it's happening in living rooms and basements, on PCs and television screens. This is the Sleeper Curve: The most debased forms of mass diversion—video games and violent television dramas and juvenile sitcoms—turn out to be nutritional after all. For decades, we've worked under the assumption that mass culture follows a steadily declining path toward lowest-common-denominator standards, presumably because the "masses" want dumb, simple pleasures and big media companies want to give the masses what they want. But in fact, the exact opposite is happening: the culture is getting more intellectually demanding, not less.

Most of the time, criticism that takes pop culture seriously involves performing some kind of symbolic analysis, decoding the work to demonstrate the way it represents some other aspect of society. You can see this symbolic approach at work in academic cultural studies programs analyzing the ways in which pop forms expressed the struggle of various disenfranchised groups: gays and lesbians, people of color, women, the third world. You can see it at work in the "zeitgeist" criticism featured in media sections of newspapers and newsweeklies, where the critic establishes a symbolic relationship between the work and some spirit of the age: yuppie self-indulgence, say, or post-9/11 anxiety.

The approach followed in this book is more systemic than symbolic, more about causal relationships than metaphors. It is closer, in a sense, to physics than to poetry. My argument for the existence of the Sleeper Curve comes out of an assumption that the landscape of popular culture involves the clash of competing forces: the neurological appetites of the brain, the economics of the culture industry, changing technological platforms. The specific ways in which those forces collide play a determining role in the type of popular culture we ultimately consume. The work of the critic, in this instance, is to diagram those forces, not decode them.

Sometimes, for the sake of argument, I find it helpful to imagine culture as a kind of man-made weather system. Float a mass of warm, humid air over cold ocean water, and you'll create an environment in which fog will thrive. The fog doesn't appear because it somehow symbolically reenacts the clash of warm air and cool water. Fog arrives instead as an emergent effect of that particular system and its internal dynamics. The same goes with popular culture: certain kinds of environments encourage cognitive complexity; others discourage complexity. The cultural object—the film or the video game—is not a metaphor for that system; it's more like an output or a result.

The forces at work in these systems operate on multiple levels: underlying changes in technology that enable new kinds of entertainment; new forms of online communications that cultivate audience commentary about works of pop culture; changes in the economics of the culture industry that encourage repeat viewing; and deep-seated appetites in the human brain that seek out reward and intellectual challenge. To understand those forces we'll need to draw upon disciplines that don't usually interact with one another: economics, narrative theory, social network analysis, neuroscience.

This is a story of trends, not absolutes. I do not believe that most of today's pop culture is made up of masterpieces that will someday be taught alongside Joyce and Chaucer in college survey courses. The television shows and video games and movies that we'll look at in the coming pages are not, for the most part, Great Works of Art. But they are more complex and nuanced than the shows and games that preceded them. While the Sleeper Curve maps *average* changes across the pop cultural landscape—and not just the complexity of single works—I have focused on a handful of representative examples in the interest of clarity. (The endnotes offer a broader survey.)

I believe that the Sleeper Curve is the single most important new force altering the mental development of young people today, and I believe it is largely a force for good: enhancing our cognitive faculties, not dumbing them down. And yet you almost never hear this story in popular accounts of today's media. Instead, you hear dire stories of addiction, violence, mindless escapism. "All across the political spectrum," television legend Steve Allen writes in a *Wall Street Journal* op-ed, "thoughtful observers are appalled by what passes for TV entertainment these days. No one can claim that the warning cries are simply the exaggerations of conservative spoil-sports or fundamentalist preachers. . . . The sleaze and classless garbage on TV in recent years exceeds the boundaries of what has traditionally been referred to as Going Too Far." The influential Parents Television Council argues: "The entertainment industry has pushed the content envelope too far; television and films filled with sex, violence, and profanity send strong negative messages to the youth of America—messages that will desensitize them and make for a far more disenfranchised society as these youths grow into adults." And then there's syndicated columnist Suzanne Fields: "The television sitcom is emblematic of our culture; parents, no matter what their degree of education, have abandoned the simplest standard of shame. Their children literally 'do not know better.' The drip, drip, drip of the popular culture dulls our senses. An open society with high technology exposes increasing numbers of adults and children to the lowest common denomination of sex and violence." You could fill an encyclopedia volume with all the kindred essays published in the past decade.

Exceptions to this dire assessment exist, but they are of the rule-proving variety. You'll see the occasional grudging acknowledgments of minor silver linings: an article will suggest that video games enhance visual memory skills, or a critic will hail *The West Wing* as the rare flowering of thoughtful programming in the junkyard of prime-time television. But the dominant motif is one of decline and atrophy: we're a nation of reality program addicts and Nintendo freaks. Lost in that account is the most interesting trend of all: that the popular culture has been growing increasingly complex over the past few decades, exercising our minds in powerful new ways.

But to see the virtue in this form of positive brainwashing, we need to begin by doing away with the tyranny of the morality play. When most op-ed writers and talk show hosts discuss the social value of media, when they address the question of whether today's media is or isn't good for us, the underlying assumption is that entertainment improves us when it carries a healthy message. Shows that promote smoking or gratuitous violence are bad for us, while those that thunder against teen pregnancy or intolerance have a positive role in society. Judged by that morality play standard, the story of popular culture over the past fifty years— if not five hundred—is a story of steady decline: the morals of the stories have grown darker and more ambiguous, and the anti-heroes have multiplied.

The usual counterargument here is that what media has lost in moral clarity it has gained in realism. The real world doesn't come in nicely packaged public service announcements, and we're better off with entertainment that reflects that fallen state with all its ethical ambiguity. I happen to be sympathetic to that argument, but it's not the one I want to make here. I think there is another way to assess the social virtue of pop culture, one that looks at media as a kind of cognitive workout, not as a series of life lessons. Those dice baseball games I immersed myself in didn't contain anything resembling moral instruction, but they nonetheless gave me a set of cognitive tools that I continue to rely on, nearly thirty years later. There may indeed be more "negative messages" in the mediasphere today, as the Parents Television Council believes. But that's not the only way to evaluate whether our television shows or video games are having a positive impact. Just as important—if not *more* important—is the kind of thinking you have to do to make sense of a cultural experience. That is where the Sleeper Curve becomes visible. Today's popular culture may not be showing us the righteous path. But it is making us smarter.

PART ONE

*   *   *

The student of media soon comes to expect the new media of any period whatever to be classed as pseudo by those who acquired the patterns of earlier media, whatever they may happen to be.

—MARSHALL MCLUHAN

GAMES

YOU CAN'T GET much more conventional than the conventional wisdom that kids today would be better off spending more time reading books, and less time zoning out in front of their video games. The latest edition of *Dr. Spock*—"revised and fully expanded for a new century" as the cover reports—has this to say of video games: "The best that can be said of them is that they may help promote eye-hand coordination in children. The worst that can be said is that they sanction, and even promote aggression and violent responses to conflict. But what can be said with much greater certainty is this: most computer games are a colossal waste of time." But where reading is concerned, the advice is quite different: "I suggest you begin to foster in your children a love of reading and the printed word from the start. . . . What is important is that your child be an avid reader."

In the middle of 2004, the National Endowment for the Arts released a study that showed that reading for pleasure had declined steadily among all major American demographic groups. The writer Andrew Solomon analyzed the consequences of this shift: "People who read for pleasure are many times more likely than those who don't to visit museums and attend musical performances, almost three times as likely to perform volunteer and charity work, and almost twice as likely to attend sporting events. Readers, in other words, are active, while nonreaders—more than half the population—have settled into apathy. There is a basic social divide between those for whom life is an accrual of fresh experience and knowledge, and those for whom maturity is a process of mental atrophy. The shift toward the latter category is frightening."

The intellectual nourishment of reading books is so deeply ingrained in our assumptions that it's hard to contemplate a different viewpoint. But as McLuhan famously observed, the problem with judging new cultural systems on their own terms is that the presence of the recent past inevitably colors your vision of the emerging form, highlighting the flaws and imperfections. Games have historically suffered from this syndrome, largely because they have been contrasted with the older conventions of reading. To get around these prejudices, try this thought experiment. Imagine an alternate world identical to ours save one techno-historical change: video games were invented and popularized *before* books. In this parallel universe, kids have been playing games for centuries—and then these page-bound texts come along and suddenly they're all the rage. What would the teachers, and the parents, and the cultural authorities have to say about this frenzy of reading? I suspect it would sound something like this:

> Reading books chronically understimulates the senses. Unlike the longstanding tradition of gameplaying—which engages the child in a vivid, three-dimensional world filled with moving images and musical soundscapes, navigated and controlled with complex muscular movements— books are simply a barren string of words on the page. Only a small portion of the brain devoted to processing written language is activated during reading, while games engage the full range of the sensory and motor cortices.
>
> Books are also tragically isolating. While games have for many years engaged the young in complex social relationships with their peers, building and exploring worlds together, books force the child to sequester him or herself in a quiet space, shut off from interaction with other children. These new "libraries" that have arisen in recent years to facilitate reading activities are a frightening sight: dozens of young children, normally so vivacious

and socially interactive, sitting alone in cubicles, reading silently, oblivious to their peers.

Many children enjoy reading books, of course, and no doubt some of the flights of fancy conveyed by reading have their escapist merits. But for a sizable percentage of the population, books are downright discriminatory. The reading craze of recent years cruelly taunts the 10 million Americans who suffer from dyslexia—a condition that didn't even exist as a condition until printed text came along to stigmatize its sufferers.

But perhaps the most dangerous property of these books is the fact that they follow a fixed linear path. You can't control their narratives in any fashion—you simply sit back and have the story dictated to you. For those of us raised on interactive narratives, this property may seem astonishing. Why would anyone want to embark on an adventure utterly choreographed by another person? But today's generation embarks on such adventures millions of times a day. This risks instilling a general passivity in our children, making them feel as though they're powerless to change their circumstances. Reading is not an active, participatory process; it's a submissive one. The book readers of the younger generation are learning to "follow the plot" instead of learning to lead.

It should probably go without saying, but it probably goes better with saying, that I don't agree with this argument. But neither is it exactly right to say that its contentions are untrue. The argument relies on a kind of amplified selectivity: it foregrounds certain isolated properties of books, and then projects worst-case scenarios based on these properties and their potential effects on the "younger generation." But it doesn't bring up any of the clear benefits of reading: the complexity of argument and storytelling offered by the book form; the stretching of the imagination triggered by reading words on a page; the shared experience you get when everyone is reading the same story.

A comparable sleight of hand is at work anytime you hear someone bemoaning today's video game obsessions, and their stupefying effects on tomorrow's generations. Games are not novels, and the ways in which they harbor novelistic aspirations are invariably the least interesting thing about them. You can judge games by the criteria designed to evaluate novels: Are the characters believable? Is the dialogue complex? But inevitably, the games will come up wanting. Games are good at novelistic storytelling the way Michael Jordan was good at playing baseball. Both could probably make a living at it, but their world-class talents lie elsewhere.

Before we get to those talents, let me say a few words about the virtues of reading books. For the record, I think those virtues are immense ones—and not just because I make a living writing books. We should all encourage our kids to read more, to develop a comfort with and an appetite for reading. But even the most avid reader in this culture is invariably going to spend his or her time with other media—with games, television, movies, or the Internet. And these other forms of culture have intellectual or cognitive virtues in their own right—different from, but comparable to, the rewards of reading.

What are the rewards of reading, exactly? Broadly speaking, they fall into two categories: the information conveyed by the book, and the mental work you have to do to process and store that information. Think of this as the difference between acquiring information and exercising the mind. When we encourage kids to read for pleasure, we're generally doing so because of the mental exercise involved. In Andrew Solomon's words: "[Reading] requires effort, concentration, attention. In exchange, it offers the stimulus to and the fruit of thought and feeling." Spock says: "Unlike most amusements, reading is an activity requiring active participation. We must do the reading ourselves—actively scan the letters, make sense of the words, and follow the thread of the story." Most tributes to the mental benefits of reading also invoke the power of imagination; reading books forces you to concoct entire worlds in your head, rather than simply ingest a series of prepackaged images. And then there is the slightly circular—though undoubtedly true—argument for the long-term career benefits: being an avid reader is good for you because the educational system and the job market put a high premium on reading skills.

To summarize, the cognitive benefits of reading involve these faculties: effort, concentration, attention, the ability to make sense of words, to follow narrative threads, to sculpt imagined worlds out of mere sentences on the page. Those benefits are themselves amplified by the fact that society places a substantial emphasis on precisely this set of skills.

The very fact that I am presenting this argument to you in the form of a book and not a television drama or a video game should make it clear that I believe the printed word remains the most powerful vehicle for conveying complicated information—though the *electronic* word is starting to give printed books a run for their money. The argument that follows is centered squarely on the side of mental exercise—and not content. I aim to persuade you of two things:

1. By almost all the standards we use to measure reading's cognitive benefits—attention, memory, following threads, and so on—the nonliterary popular culture has been steadily growing more challenging over the past thirty years.

2. Increasingly, the nonliterary popular culture is honing *different* mental skills that are just as important as the ones exercised by reading books.

Despite the warnings of Dr. Spock, the most powerful examples of both these trends are found in the world of video games. Over the past few years, you may have noticed the appearance of a certain type of story about gaming culture in mainstream newspapers and periodicals. The message of that story ultimately reduces down to: Playing video games may not actually be a *complete* waste of time. Invariably these stories point to some new study focused on a minor side effect of gameplaying—often manual dexterity or visual memory—and explain that heavy gamers show improved skills compared to non-gamers. (The other common let's-take-games-seriously story is financial, usually pointing to the fact that the gaming industry now pulls in more money than Hollywood.)

Now, I have no doubt that playing today's games does in fact improve your visual intelligence and your manual dexterity, but the virtues of gaming run far deeper than hand-eye coordination. When I read these ostensibly positive accounts of video games, they strike me as the equivalent of writing a story about the merits of the great novels and focusing on how reading them can improve your spelling. It's true enough, I suppose, but it doesn't do justice to the rich, textured experience of novel reading. There's a comparable blindness at work in the way games have been covered to date. For all the discussion of gaming culture that you see, the actual experience of playing games has been strangely misrepresented. We hear a lot about the content of games:

the carnage and drive-by killings and adolescent fantasies. But we rarely hear accurate descriptions about what it actually *feels like* to spend time in these virtual worlds. I worry about the experiential gap between people who have immersed themselves in games, and people who have only heard secondhand reports, because the gap makes it difficult to discuss the meaning of games in a coherent way. It reminds me of the way the social critic Jane Jacobs felt about the thriving urban neighborhoods she documented in the sixties: "People who know well such animated city streets will know how it is. People who do not will always have it a little wrong in their heads—like the old prints of rhinoceroses made from travelers' descriptions of the rhinoceroses."

So what does the rhinoceros actually look like? The first and last thing that should be said about the experience of playing today's video games, the thing you almost never hear in the mainstream coverage, is that games are fiendishly, sometimes maddeningly, *hard*.

*   *   *

THE DIRTY little secret of gaming is how much time you spend not having fun. You may be frustrated; you may be confused or disoriented; you may be stuck. When you put the game down and move back into the real world, you may find yourself mentally working through the problem you've been wrestling with, as though you were worrying a loose tooth. If this is mindless escapism, it's a strangely masochistic version. Who wants to escape to a world that irritates you 90 percent of the time?

Consider the story of Troy Stolle, a construction site worker from Indianapolis profiled by the technology critic Julian Dibbell. When he's not performing his day job as a carpenter building wooden molds, Stolle lives in the virtual world of *Ultima Online,* the fantasy-themed game that allows you to create a character—sometimes called an avatar—and interact with thousands of other avatars controlled by other humans, connected to the game over the Net. (Imagine a version of Dungeons & Dragons where you're playing with thousands of strangers from all over the world, and you'll get the idea.) *Ultima* and related games like *EverQuest* have famously developed vibrant simulated economies that have begun to leak out into the real world. You can buy a magic sword or a plot of land—entirely made of digital code, mind you—for hundreds of dollars on eBay. But earning these goods the old-fashioned within-the-gameworld way takes time—a lot of time. Dibbell describes the ordeal Stolle had to go through to have his avatar, named Nils Hansen, purchase a new house in the *Ultima* world:

> Stolle had had to come up with the money for the deed. To get the money, he had to sell his old house. To get that house in the first place, he had to spend hours crafting vir-

tual swords and plate mail to sell to a steady clientele of about three dozen fellow players. To attract and keep that clientele, he had to bring Nils Hansen's blacksmithing skills up to Grandmaster. To reach that level, Stolle spent six months doing nothing but smithing: He clicked on hillsides to mine ore, headed to a forge to click the ore into ingots, clicked again to turn the ingots into weapons and armor, and then headed back to the hills to start all over again, each time raising Nils' skill level some tiny fraction of a percentage point, inching him closer to the distant goal of 100 points and the illustrious title of Grandmaster Blacksmith.

Take a moment now to pause, step back, and consider just what was going on here: Every day, month after month, a man was coming home from a full day of bone-jarringly repetitive work with hammer and nails to put in a full night of finger-numbingly repetitive work with "hammer" and "anvil"—and paying $9.95 per month for the privilege. Ask Stolle to make sense of this, and he has a ready answer: "Well, it's not work if you enjoy it." Which, of course, begs the question: Why would anyone enjoy it?

Why? Anyone who has spent more than a few hours trying to complete a game knows the feeling: you get to a point where there's a sequence of tasks you know you have to complete to proceed further into the world, but the tasks themselves are more like chores than entertainment, something you *have* to do, not something you want to do: building roads and laying power lines, retreating through a tunnel sequence to find an object you've left behind, conversing with characters when you've already memorized their lines. And yet a large part of the population performing these tasks every day is composed of precisely the demographic group most averse to doing chores. If you practically have to lock kids in their room to get them to do their math homework, and threaten to ground them to get them to take out the trash, then why are they willing to spend six months smithing in *Ultima*? You'll often hear video games included on the list of the debased instant gratifications that abound in our culture, right up there with raunchy music videos and fast food. But compared to most forms of popular entertainment, games turn out to be all about *delayed* gratification—sometimes so long delayed that you wonder if the gratification is ever going to show.

The clearest measure of the cognitive challenges posed by modern games is the sheer size of the cottage industry devoted to publishing game guides, sometimes called walkthroughs, that give you detailed, step-by-step explanations of how to complete the game that is currently torturing you. During my twenties, I'd wager that I spent somewhere shockingly close to a thousand dollars buying assorted cheat sheets, maps, help books, and phone support to assist my usually futile attempt to complete a video game. My relationship to these reference texts is intimately bound up with my memory of each game, so that the *Myst* sequel *Riven* brings to mind those hours on the automated phone support line, listening to a recorded voice explain that the lever has to be rotated 270 degrees before the blue pipe will connect with the transom, while the playful *Banjo-Kazooie* conjures up a cheery atlas of vibrant level maps, like a child's book where the story has been replaced with linear instruction sets: jump twice on the mushroom, then grab the gold medallion in the moat. Admitting just how much money I spent on these guides sounds like a cry for help, I know, but the great, looming racks of these game guides at most software stores are clear evidence that I am not alone in this habit. The guidebook for the controversial hit game *Grand Theft Auto* alone has sold more than 1.6 million copies.

Think about the existence of these guides in the context of other forms of popular entertainment. There are plenty of supplementary texts that accompany Hollywood movies or Billboard chart-toppers: celebrity profiles, lyrics sheets, reviews, fan sites, commentary tracks on DVDs. These texts can widen your understanding of a film or an album, but you'll almost never find yourself *needing* one. People don't walk into theaters with guidebooks that they consult via flashlight during the film. But they regularly rely on these guides when playing a game. The closest cultural form to the game guide is the august tradition of CliffsNotes marketed as readers' supplements to the Great Books. There's nothing puzzling about the existence of CliffsNotes: we accept both the fact that the Great Books are complicated, and the fact that millions of young people are forced more or less against their will to at least pretend to read them. Ergo: a thriving market for CliffsNotes. Game guides, however, confound our expectations: because we're not used to accepting the complexity of gaming culture, and because nobody's forcing the kids to master these games.

The need for such guides is a relatively new development: you didn't need ten pages to explain the *PacMan* system, but two hundred pages barely does justice to an expanding universe like *EverQuest* or *Ultima*. You need them because the complexity of these worlds can be overwhelming: you're stuck in the middle of a level, with all the various exits locked and no sign of a key. Or the password for the control room you thought you found two hours ago turns out not to work. Or the worst case: you're wandering aimlessly through hallways, like those famous tracking shots from *The Shining,* and you've got no real idea what you're supposed to be doing next.

This aimlessness, of course, is the price of interactivity. You're more in control of the narrative now, but your supply of information about the narrative—whom you should talk to next, where that mysterious package has been hidden—is only partial, and so playing one of these games is ultimately all about filling in that information gap. When it works, it can be exhilarating, but when it doesn't—well, that's when you start shelling out the fifteen bucks for the cheat sheet. And then you find yourself hunched over the computer screen, help guide splayed open on the desk, flipping back and forth between the virtual world and the level maps, trying to find your way. After a certain point—perhaps when the level maps don't turn out to be all that helpful, or perhaps when you find yourself reading the help guides over dinner—you start saying to yourself: Remind me why this is fun?

*   *   *

So why does anyone bother playing these things? Why do we use the word "play" to describe this torture? I'm always amazed to see what our brains are willing to tolerate to reach the next level in these games. Several years ago I found myself on a family vacation with my seven-year-old nephew, and on one rainy day I decided to introduce him to the wonders of *SimCity 2000*, the legendary city simulator that allows you to play Robert Moses to a growing virtual metropolis. For most of our session, I was controlling the game, pointing out landmarks as I scrolled around my little town. I suspect I was a somewhat condescending guide—treating the virtual world as more of a model train layout than a complex system. But he was picking up the game's inner logic nonetheless. After about an hour of tinkering, I was concentrating on trying to revive one particularly run-down manufacturing district. As I contemplated my options, my nephew piped up: "I think we need to lower our industrial tax rates." He said it as naturally, and as confidently, as he might have said, "I think we need to shoot the bad guy."

The interesting question here for me is not whether games are, on the whole, more complex than most other cultural experiences targeted at kids today—I think the answer to that is an emphatic yes. The question is why kids are so eager to soak up that much information when it is delivered to them in game form. My nephew would be asleep in five seconds if you popped him down in an urban studies classroom, but somehow an hour of playing *SimCity* taught him that high tax rates in industrial areas can stifle development. That's a powerful learning experience, for reasons we'll explore in the coming pages. But let's start with the more elemental question of desire. Why does a seven-year-old soak up the intricacies of industrial economics in game form, when the same subject would send him screaming for the exits in a classroom?

The quick explanations of this mystery are not helpful. Some might say it's the flashy graphics, but games have been ensnaring our attention since the days of *Pong,* which was—graphically speaking—a huge step backward compared with television or movies, not to mention reality. Others would say it's the violence and sex, and yet games like *SimCity*—and indeed most of the best-selling games of all time—have almost no violence and sex in them. Some might argue that it's the interactivity that hooks, the engagement of building your own narrative. But if active participation alone functions as a drug that entices the mind, then why isn't the supremely *passive* medium of television repellant to kids?

Why do games captivate? I believe the answer involves a deeper property that most games share—a property that will be instantly familiar to anyone who has spent time in this world, but one that is also strangely absent from most outside descriptions. To appreciate this property you need to look at game culture through the lens of neuroscience. There's a logical reason to use that lens, of course: If you're trying to figure out why cocaine is addictive, you need a working model of what cocaine is, and you need a working model of how the brain functions. The same goes for the question of why games are such powerful attractors. Explaining that phenomenon without a working model of the mind tells only half the story.

This emphasis on the inner life of the brain will be a recurring theme in the coming pages. Cultural critics like to speculate on the cognitive changes induced by new forms of media, but they rarely invoke the insights of brain science and other empirical research in backing up those claims. All too often, this has the effect of reducing their arguments to mere superstition. If you're trying to make sense of a new cultural form's effect on the way we view the world, you need to be able to describe the cultural object in some detail, and also demonstrate how that object transforms the mind that is apprehending it. In some instances, you can measure that transformation through traditional modes of intelligence testing; in some cases, you can measure changes by looking at brain activity directly, thanks to modern scanning technology; and in cases where the empirical research hasn't yet been done, you can make informed speculation based on our understanding of how the brain works.

To date, there has been very little direct research into the question of how games manage to get kids to learn without realizing that they're learning. But a strong case can be made that the power of games to captivate involves their ability to tap into the brain's natural reward circuitry. Because of its central role in drug addiction, the reward circuits of the brain have been extensively studied and mapped in recent years. Two insights that have emerged from this study are pertinent to the understanding of games. First, neuroscientists have drawn a crucial distinction between the way the brain seeks out reward and the way it delivers pleasure. The body's natural painkillers, the opioids, are the brain's pure pleasure drugs, while the reward system revolves around the neurotransmitter dopamine interacting with specific receptors in a part of the brain called the nucleus accumbens.

The dopamine system is a kind of accountant: keeping track of expected rewards, and sending out an alert—in the form of lowered dopamine levels—when those rewards don't arrive as promised. When the pack-a-day smoker deprives himself of his morning cigarette; when the hotshot Wall Street trader doesn't get the bonus he was planning on; when the late-night snacker opens the freezer to find someone's pilfered all the Ben & Jerry's—the disappointment and craving these people experience is triggered by lowered dopamine levels.

The neuroscientist Jaak Panksepp calls the dopamine system the brain's "seeking" circuitry, propelling us to seek out new avenues for reward in our environment. Where our brain wiring is concerned, the craving instinct triggers a desire to explore. The system says, in effect: "Can't find the reward you were promised? Perhaps if you just look a little harder you'll be in luck—it's got to be around here somewhere."

How do these findings connect to games? Researchers have long suspected that geometric games like *Tetris* have such a hypnotic hold over us (longtime *Tetris* players have vivid dreams about the game) because the game's elemental shapes activate modules in our visual system that execute low-level forms of pattern recognition—sensing parallel and perpendicular lines, for instance. These modules are churning away in the background all the time, but the simplified graphics of *Tetris* bring them front and center in our consciousness. I believe that what *Tetris* does to our visual circuitry, most video games do to the reward circuitry of the brain.

Real life is full of rewards, which is one reason why there

are now so many forms of addiction. You can be rewarded by love and social connection, financial success, drug abuse, shopping, chocolate, and watching your favorite team win the Super Bowl. But supermarkets and shopping malls aside, most of life goes by without the potential rewards available to you being clearly defined. You know you'd like that promotion, but it's a long way off, and right now you've got to deal with getting this memo out the door. Real-life reward usually hovers at the margins of day-to-day existence—except for the more primal rewards of eating and making love, both of which exceed video games in their addictiveness.

In the gameworld, reward is everywhere. The universe is literally teeming with objects that deliver very clearly articulated rewards: more life, access to new levels, new equipment, new spells. Game rewards are fractal; each scale contains its own reward network, whether you're just learning to use the controller, or simply trying to solve a puzzle to raise some extra cash, or attempting to complete the game's ultimate mission. Most of the crucial work in game interface design revolves around keeping players notified of potential rewards available to them, and how much those rewards are currently needed. Just as *Tetris* streamlines the fuzzy world of visual reality to a core set of interacting shapes, most games offer a fictional world where rewards are larger, and more vivid, more clearly defined, than life. This is true even of games that have been rightly celebrated for their open-endedness. *SimCity* is famous for not forcing the player along a preordained narrative line; you can build any kind of community you want: small farming villages, vast industrial Coketowns, high-centric edge cities or pedestrian-friendly neighborhoods. But the game has a subtle reward architecture that plays a major role in the game's addictiveness: the software withholds a trove of objects and activities until you've reached certain predefined levels, either of population, money, or popularity. You can build pretty much any kind of environment you want playing *SimCity,* but you can't build a baseball stadium until you have fifty thousand residents. Similarly, *Grand Theft Auto* allows players to drive aimlessly through a vast urban environment, creating their own narratives as they explore the space. But for all that open-endedness, the game still forces you to complete a series of pre-defined missions before you are allowed to enter new areas of the city. The very games that are supposed to be emblems of unstructured user control turn out to dangle rewards at every corner.

"Seeking" is the perfect word for the drive these designs instill in their players. You want to win the game, of course, and perhaps you want to see the game's narrative completed. In the initial stages of play, you may just be dazzled by the game's graphics. But most of the time, when you're hooked on a game, what draws you in is an elemental form of desire: the desire to *see the next thing.* You want to cross that bridge to see what the east side of the city looks like, or try out that teleportation module, or build an aquarium on the harbor. To someone who has never felt that sort of compulsion, the underlying motivation can seem a little strange: you want to build the aquarium not, in the old mountaineering expression, because it's there, but rather because it's not there, or not there *yet*. It's not there, but you know—because you've read the manual or the game guide, or because the interface is flashing it in front of your eyes— you know that if you just apply yourself, if you spend a little more time cultivating new residents and watching the annual budget, the aquarium will eventually be yours to savor.

In a sense, neuroscience has offered up a prediction here, one that games obligingly confirm. If you create a system where rewards are both clearly defined and achieved by exploring an environment, you'll find human brains drawn to those systems, even if they're made up of virtual characters and simulated sidewalks. It's not the subject matter of these games that attracts—if that were the case, you'd never see twenty-somethings following absurd rescue-the-princess storylines like the best-selling *Zelda* series on the Nintendo platform. It's the reward system that draws those players in, and keeps their famously short attention spans locked on the screen. No other form of entertainment offers that cocktail of reward and exploration: we don't "explore" movies or television or music in anything but the most figurative sense of the word. And while there are rewards to those other forms—music in fact has been shown to trigger opioid release in the brain—they don't come in the exaggerated, tantalizing packaging that video games wrap around them.

You might reasonably object at this point that I have merely demonstrated that video games are the digital equivalent of crack cocaine. Crack also has a powerful hold over the human brain, thanks in part to its manipulations of the dopamine system. But that doesn't make it a good thing. If games have been unwittingly designed to lock into our brain's reward architecture, then what positive value are we getting out of that intoxication? Without that positive value the Sleeper Curve is meaningless.

Here again, you have to shed your expectations about older cultural forms to make sense of the new. Game players are not soaking up moral counsel, life lessons, or rich psychological portraits. They are not having emotional experiences with their Xbox, other than the occasional adrenaline rush. The narratives they help create now rival pulp Hollywood fare, which is an accomplishment when measured against the narratives of *PacMan* and *Pong,* but it's still setting the bar pretty low. With the occasional exception, the actual *content* of the game is often childish or gratuitously menacing—though, again, not any more so than your average summer blockbuster. Complex social and historical simulations like *Age of Empires or Civilization* do dominate the game charts, and no doubt these games do impart some useful information about ancient Rome or the design of mass transit systems. But much of the roleplay inside the gaming world alternates between drive-by shooting and princess rescuing.

De-emphasizing the content of game culture shouldn't be seen as a cop-out. We ignore the content of many activities that are widely considered to be good for the brain or the body. No one complains about the simplistic, militaristic plot of chess games. ("It always ends the same way!") We teach algebra to children knowing full well that the day they leave the classroom, ninety-nine percent of those kids will never again directly employ their algebraic skills. Learning algebra isn't about acquiring a specific tool; it's about building up a mental muscle that will come in handy elsewhere. You don't go to the gym because you're interested in learning how to operate a StairMaster; you go to the gym because operating a StairMaster does something laudable to your body, the benefits of which you enjoy during the many hours of the week when you're not on a StairMaster.

So it is with games. It's not *what* you're thinking about when you're playing a game, it's *the way* you're thinking that matters. The distinction is not exclusive to games, of course. Here's John Dewey, in his book *Experience and Education*: "Perhaps the greatest of all pedagogical fallacies is the notion that a person learns only that particular thing he is studying at the time. Collateral learning in the way of formation of enduring attitudes, of likes and dislikes, may be and often is much more important than the spelling lesson or lesson in geography or history that is learned. For these attitudes are fundamentally what count in the future."

This is precisely where we need to make our portrait of the rhinoceros as accurate as possible: defining the collateral learning that goes beyond the explicit content of the experience. Start with the basics: far more than books or movies or music, games force you to make *decisions*. Novels may activate our imagination, and music may conjure up powerful emotions, but games force you to decide, to choose, to prioritize. All the intellectual benefits of gaming derive from this fundamental virtue, because learning how to think is ultimately about learning to make the right decisions: weighing evidence, analyzing situations, consulting your long-term goals, and then deciding. No other pop cultural form directly engages the brain's decision-making apparatus in the same way. From the outside, the primary activity of a gamer looks like a fury of clicking and shooting, which is why so much of the conventional wisdom about games focuses on hand-eye coordination. But if you peer inside the gamer's mind, the primary activity turns out to be another creature altogether: making decisions, some of them snap judgments, some long-term strategies.

Those decisions are themselves predicated on two modes of intellectual labor that are key to the collateral learning of playing games. I call them *probing* and *telescoping*.

❖   ❖   ❖

MOST VIDEO GAMES differ from traditional games like chess or Monopoly in the way they withhold information about the underlying rules of the system. When you play chess at anything beyond a beginner's level, the rules of the game contain no ambiguity: you know exactly the moves allowed for each piece, the procedures that allow one piece to capture another. The question that confronts you sitting down at the chessboard is not: What are the rules here? The question is: What kind of strategy can I concoct that will best exploit those rules to my advantage?

In the video game world, on the other hand, the rules are rarely established in their entirety before you sit down to play. You're given a few basic instructions about how to manipulate objects or characters on the screen, and a sense of some kind of immediate objective. But many of the rules—the identity of your ultimate goal and the techniques available for reaching that goal—become apparent only through exploring the world. You literally learn by playing. This is one reason video games can be frustrating to the noninitiated. You sit down at the computer and say, "What am I supposed to do?" The regular gamers in the room have to explain: "You're supposed to figure out what you're supposed to do." You have to probe the depths of the game's logic to make sense of it, and like most probing expeditions, you get results by trial and error, by stumbling across things, by following hunches. In almost every other endeavor that we describe using the language of games—poker, baseball, backgammon, capture the flag—any ambiguity in the rules and objectives of the game would be a fatal flaw. In video games, on the other hand, it's a core part of the experience. Many game narratives contain mysteries of sorts modeled after Hollywood plotlines—Who murdered my brother? Who stole the plutonium?—but the ultimate mystery that drives players deeper into the gameworld is a more self-referential one: how is this game played? Non-gamers usually imagine that mastering a game is largely a matter of learning to push buttons faster, which no doubt accounts for all the "hand-eye coordination" clichés. But for many popular games, the ultimate key to success lies in deciphering the rules, and not manipulating joysticks.

Probing involves a nuanced form of exploration as well, one that often operates below conscious awareness. Video games obviously differ from traditional games like chess or basketball in that the entire game environment is created by a computer. Explicit rules are a crucial part of that environment: you learn that you have only three lives, or that you can't build a marina until you have fifty thousand residents, or that you can't open the gate on the third level until you find the key on the second. Some of these rules you can learn just by reading the manual; others have to be discovered by playing. But the computer is doing more than just serving up clearly defined rules; it's concocting an entire world, a world with biology, light, economies, social relations, weather. I call this the *physics* of the virtual world—as opposed to the rules of the game—though this kind of physics goes well beyond acceleration curves and gravity.

You're probing the physics of a world when you start detecting subtle patterns and tendencies in the way the computer is running the simulation. Sometimes these have to do with mass and velocity: you can't jump across the canyon if you're wearing your armor; the rocket launcher is the only weapon that can shoot far enough to attack from the rear of the fortress. Sometimes they have to do with physiology: you'll lose more blood if you're wounded in the chest than in the legs; you can jump from any height without injuring your character. Sometimes it's collective behavior: your neighbors stay longer at the party if you have a jukebox and a Lava lamp; the invading robots tend to swoop in from the right when you first land on the planet. When my nephew suggested lowering the industrial tax rate during my demo of *SimCity*, he was probing the game's physics. I had explained the official rules to him: players are allowed to alter the tax rates for different zones. The physics were fuzzier, more intuitive: if you lower the rate in a given area, you'll usually see some growth there, assuming the other variables—power, water, crime—aren't impeding development.

The game scholar James Paul Gee breaks probing down

into a four-part process, which he calls the "probe, hypothesize, reprobe, rethink" cycle:

1. The player must *probe* the virtual world (which involves looking around the current environment, clicking on something, or engaging in a certain action).

2. Based on reflection while probing and afterward, the player must form a *hypothesis* about what something (a text, object, artifact, event, or action) might mean in a usefully situated way.

3. The player *reprobes* the world with that hypothesis in mind, seeing what effect he or she gets.

4. The player treats this effect as feedback from the world and accepts or *rethinks* his or her original hypothesis.

Put another way: When gamers interact with these environments, they are learning the basic procedure of the scientific method.

Probing often takes the form of seeking out the limits of the simulation, the points at which the illusion of reality breaks down, and you can sense that's all just a bunch of algorithms behind the curtain. The first celebrated instance of this arrived in the early eighties with the hugely popular arcade game *PacMan*. The game had its rules, which were so simple you could express them in three sentences: gobble all the dots to finish a level; avoid the monsters unless you've eaten one of the large dots, at which point you can eat the monsters; eat the prizes for extra points. But experienced *PacMan* players soon discovered that the monsters roamed the maze in predictable ways, and if you followed a certain course—literally called a "pattern"—you'd complete the level without losing a man every time you played. Patterns weren't built into the official rules of the game; they were a legacy effect of the limited computational power of those arcade machines, and the predictable way in which the monsters' behavior had been programmed. To detect those limitations, you had to probe the *PacMan* game by playing it hundreds of times, experimenting with different strategies until one sequence revealed itself.

Probing the limits of the game physics is another oft-ignored facet of gaming culture. I suspect most hard-core gamers would acknowledge that part of the pleasure of their immersions comes from this kind of pursuit, searching out the points where the system shows its flaws— partially because those flaws can be exploited, as in *PacMan*'s patterns, but also because there's something strangely satisfying about defining the edges of a simulation, learning what it's capable of and where it breaks down. Some people find this kind of exploration appealing in ordinary life: they're the sort that actually enjoys looking under the hood of the car, or memorizing UNIX commands. But video games *force* you to speculate about what's going on under the hood. If you don't think about the underlying mechanics of the simulation—even if that thinking happens in a semiconscious way—you won't last very long in the game. You have to probe to progress.

I didn't have a word for it at the time, of course, but I now realize that my tour through the universe of dice-baseball was a way of probing the physics of those early games. I'd learn the explicit rules for each simulation, but the really fascinating moment came when I'd start rolling the dice and generating results. Only by playing the simulations could you get a sense of their realism. Usually, you had to work through a quarter of a season before the imperfections would reveal themselves: batters would strike out too frequently in one simulation; another would allow sluggers to average an implausible two home runs a game. I was detecting flaws in these systems, but there was nonetheless something profoundly satisfying about the experience. Bringing these imperfections to light felt like solving a mystery, looking past the surface illusion of player cards and charts to the inner truth of the system.

*   *   *

ONE OF the best ways to grasp the cognitive virtues of gameplaying is to ask committed players to describe what's going on in their heads halfway through a long virtual adventure like *Zelda* or *Half-Life*. It's crucial here not to ask what's happening in the gameworld, but rather what's happening to the players mentally: what problems they're actively working on, what objectives they're trying to achieve. In my experience, most gamers will be more inclined to show rather than tell the probing they've done; they'll have internalized flaws or patterns in the simulation without being fully aware of what they're doing. Certain strategies just *feel* right.

But if the gamers' probing is semiconscious, their awareness of mid-game *objectives* will be crystal clear. They'll be able to give you an explicit account of what they need to do to reach the goals that the game has laid out for them. Many of these goals will have been obscure in the opening sequences of the game, but by the halfway point, players have usually constructed a kind of to-do list that governs their strategy. If probing is all about depth, exploring the buried logic of the simulation, tracking objectives is a kind of temporal thinking, a looking forward to all the hurdles that separate you from the game's completion.

Tracking objectives seems simple enough. If you stopped playing in the early nineties, or if you only know about games from secondhand accounts, you'd probably assume that the mid-game objectives would sound something like this: Shoot that guy over there! Or: Avoid the blue monsters! Or: Find the magic key!

But interrupt a player in the middle of a *Zelda* quest, and ask her what her objectives are, and you'll get a much more interesting answer. Interesting for two reasons: first, the sheer number of objectives simultaneously at play; and second, the nested, hierarchical way in which those objectives have to be mentally organized. For comparison's sake, here's what the state of mind of a *PacMan* player would look like mid-game circa 1981:

1. Move the joystick in order to . . .
   2. Eat all the dots in order to . . .
      3. Get to the next level in order to . . .
         4. Reach level 256 (the final one) or a new
            high score.

Those objectives could be mildly complicated with the addition of one subcategory, which would look like this:

1. Your ultimate goal is to clear all the boards of dots.
   2. Your immediate goal is to complete the
      current maze.
      3. To do this, you must move the joystick through
         the maze and avoid the monsters.
         3a. You may also clear the board of monsters
            by eating large dots.
            3b. You may also eat the fruit for bonus
               points.

A real-world game like checkers would generate a list of comparable simplicity:

1. Your goal is to capture all of your opponent's pieces.
2. To do this, you must move one piece each turn, capturing pieces where possible.
   2a. You may also revive your own captured pieces by reaching the other side of the board.

A map of the objectives in the latest *Zelda* game, *The Wind Walker*, looks quite different:

1. Your ultimate goal is to rescue your sister.
2. To do this, you must defeat the villain Ganon.
3. To do this, you need to obtain legendary weapons.
4. To locate the weapons, you need the pearl of Din.
5. To get the pearl of Din, you need to cross the ocean.
6. To cross the ocean, you need to find a sailboat.
7. To do all the above, you need to stay alive and healthy.
8. To do all the above, you need to move the controller.

The eight items can be divided into two groups, each with a slightly different purchase on the immediate present. The last two items (7 and 8) are almost metabolic in nature, the basics of virtual self-preservation: keep your character alive, with maximum power and, where possible, flush with cash. Like many core survival behaviors, some of these objectives take quite a bit of training—learning the navigation interface and mapping it onto the controller, for instance—but once you've mastered them, you don't necessarily have to think about what you're doing. You've internalized or automated the knowledge, just as you did years ago when you learned how to run or climb or talk.

Beyond the horizon of those immediate needs lie the six remaining master objectives. These are forward projections that color the immediate present. They're like constellations guiding your ship through uncertain waters. Lose sight of them and you're adrift.

But those master objectives are rarely the player's central focal point, because most of the game is spent solving smaller problems that stand in the way of achieving one of the primary goals. In this sense, our list of eight nested objectives is a gross simplification of the actual problem-solving that goes on in a game like *Zelda*. Zoom in on just one of these objectives—finding the pearl of Din—and the list of objectives running through the player's mind would look something like this:

To locate the items, you need the pearl of Din from
the islanders.
To get this, you need to help them solve
their problem.
To do this, you need to cheer up the Prince.
To do this, you need to get a letter from
the girl.
To do this, you need to find the girl in
the village.

With the letter to the Prince, you must now befriend
the Prince.
To do this, you need to get to the top of Dragon
Roost Mt.
To do this, you must get to the other side of
the gorge.
To do this, you must fill up the gorge with
water so you can swim across.
To do this, you must use a bomb to blow up
the rock blocking the water.
To do this, you must make the bomb
plant grow.
To do this, you must collect water in a
jar that the girl gave you.
Once on the other side, you must cross lava.
To do this, you must knock down statues on either
side of the lava.

To do this, you must throw bombs into holes in
the statues.
To do this, you must pull up bombs and
aim them.
Once past the lava, you must get into the cavern.
To do this, you must pull statues out of the way.
Once in the cavern, you must get to the next room.
To do this, you need to kill the guards in your way.
To do this, you need to fight with the controller.
To do this, you need to obtain a key to the
locked door.
To do this, you must light the two unlit
torches in the room.
To do this, you must obtain your own
source of fire.
To do this, you must pick up a wooden
staff and light it.

I'll spare you the entire sequence for this one objective, which would continue on for another page unabridged. And remember, this is merely a snapshot of an hour or so of play from a title that averages around forty hours to complete. And remember, too, that almost all of these objectives have to be deciphered by the player on his own, assuming he's not consulting a game guide. These local objectives make up the primary texture of the game; they're what you spend most of your time working through. Gamers sometimes talk about the units formed by these steps as a "puzzle." You hit a point in the game where you know you need to do something, but there's some obstruction in your way, and the game conventions signal to you that you've encountered a puzzle. You're not lost, or confused; in fact, you're on precisely the right track—it's just the game designers have artfully deposited a puzzle in the middle of that track.

I call the mental labor of managing all these simultaneous objectives "telescoping" because of the way the objectives nest inside one another like a collapsed telescope. I like the term as well because part of this skill lies in focusing on immediate problems while still maintaining a long-distance view. You can't progress far in a game if you simply deal with the puzzles you stumble across; you have to coordinate them with the ultimate objectives on the horizon. Talented gamers have mastered the ability to keep all these varied objectives alive in their heads simultaneously.

Telescoping should not be confused with multitasking. Holding this nested sequence of interlinked objectives in your mind is not the same as the classic multitasking teenager scenario, where they're listening to their iPod while instant messaging their friends and Googling for research on a term paper. Multitasking is the ability to handle a chaotic stream of unrelated objectives. Telescoping is all about order, not chaos; it's about constructing the proper hierarchy of tasks and moving through the tasks in the correct sequence. It's about perceiving relationships and determining priorities.

If telescoping involves a sequence, by the same token the feeling it conjures in the brain is not, I think, a *narrative* feeling. There are layers to narratives, to be sure, and they inevitably revolve around a mix of the present and future, between what's happening now and the tantalizing question of where it's all headed. But narratives are built out of events, not tasks. They happen *to* you. In the gameworld you're forced to define and execute the tasks; if your definitions get blurry or are poorly organized, you'll have trouble playing. You can still enjoy a book without explicitly concentrating on where the narrative will take you two chapters out, but in gameworlds you need that long-term planning as much as you need present-tense focus. In a sense, the closest analog to the way gamers are thinking is the way programmers think when they write code: a nested series of instructions with multiple layers, some focused on the basic tasks of getting information in and out of memory, some focused on higher-level functions like how to represent the program's activity to the user. A program is a sequence, but not a narrative; playing a video game generates a series of events that retrospectively sketch out a narrative, but the pleasures and challenges of playing don't equate with the pleasures of following a story.

There is something profoundly *lifelike* in the art of probing and telescoping. Most video games take place in worlds
that are deliberately fanciful in nature, and even the most realistic games can't compare to the vivid, detailed illusion of
reality that novels or movies concoct for us. But our lives are
not stories, at least in the present tense—we don't passively
consume a narrative thread. (We turn our lives into stories
after the fact, after the decisions have been made, and the
events have unfolded.) But we do probe new environments
for hidden rules and patterns; we do build telescoping hier-
archies of objectives that govern our lives on both micro
and macro time frames. Traditional narratives have much to
teach us, of course: they can enhance our powers of communication, and our insight into the human psyche. But if
you were designing a cultural form explicitly to train the
cognitive muscles of the brain, and you had to choose between a device that trains the mind's ability to follow narrative events, and one that enhanced the mind's skills at
probing and telescoping—well, let's just say we're fortunate
not to have to make that choice.

Still, I suspect that some readers may be cringing at the
subject matter of those *Zelda* objectives. Here again, the
problem lies in adopting aesthetic standards designed to
evaluate literature or drama in determining whether we
should take the video games seriously. Consider this sequence from our telescoping inventory:

With the letter to the Prince, you must now befriend
the Prince.
To do this, you need to get to the top of Dragon
Roost Mt.
To do this, you must get to the other side of
the gorge.
To do this, you must fill up the gorge with
water so you can swim across.
To do this, you must use a bomb to blow up
the rock blocking the water.
To do this, you must make the bomb
plant grow.
To do this, you must collect water in a
jar that the girl gave you.

If you approach this description with aesthetic expectations borrowed from the world of literature, the content
seems at face value to be child's play: blowing up bombs
to get to Dragon Roost Mountain; watering explosive
plants. A high school English teacher would look at this
and say: There's no psychological depth here, no moral
quandaries, no poetry. And he'd be right! But comparing
these games to *The Iliad* or *The Great Gatsby* or *Hamlet*
relies on a false premise: that the intelligence of these
games lies in their content, in the themes and characters
they represent. I would argue that the cognitive challenges

of videogaming are much more usefully compared to another educational genre that you will no doubt recall from your school days:

Simon is conducting a probability experiment. He randomly selects a tag from a set of tags that are numbered from 1 to 100 and then returns the tag to the set. He is trying to draw a tag that matches his favorite number, 21. He has not matched his number after 99 draws.

What are the chances he will match his number on the 100th draw?

A. 1 out of 100
B. 99 out of 100
C. 1 out of 1
D. 1 out of 2

Judged by the standards employed by our English teacher, this passage—taken from the Massachusetts Comprehensive Assessment exam for high-school math—would be an utter failure. Who is this Simon? We know nothing about him; he is a cipher to us, a prop. There are no flourishes in the prose, nothing but barren facts, describing a truly useless activity. Why would anyone want to number a hundred tags and then go about trying to randomly select a favorite number? What is Simon's motivation?

Word problems of this sort have little to offer in the way of moral lessons or psychological depth; they won't make students more effective communicators or teach them technical skills. But most of us readily agree that they are good for the mind on some fundamental level: they teach abstract skills in probability, in pattern recognition, in understanding causal relations that can be applied in countless situations, both personal and professional. The problems that confront the gamers of *Zelda* can be readily translated into this form, and indeed in translating a core property of the experience is revealed:

You need to cross a gorge to reach a valuable destination. At one end of the gorge a large rock stands in front of a river, blocking the flow of water. Around the edge of the rock a number of small flowers are growing. You have been given a jar by another character. How can you cross the gorge?

A. Jump across it.
B. Carry small pails of water from the river and pour them in the gorge, and then swim across.
C. Water the plants, and then use the bombs they grow to blow up the rock, releasing the water, and then swim across.
D. Go back and see if you've missed some important tool in an earlier scene.

Again, the least interesting thing about this text is the substance of the story. You could perhaps meditate on the dramatic irony inherent in bomb-growing flowers, or analyze the gift economy relationship introduced with the crucial donation of the jar. But those interpretations will go only so far, because what's important here is not the content of the *Zelda* world, but the way that world has been organized to tax the problem-solving skills of the player. To be sure, the pleasure of gaming goes beyond this kind of problem-solving; the objects and textures of the worlds offer rich aesthetic experiences; many networked games offer intriguing social exchanges; increasingly the artificial intelligence embedded in some virtual characters provides amazing interactions. But these are all ultimately diversions. You can't make progress in the game without learning the rules of the environment. On the simplest level, the *Zelda* player learns how to grow bombs out of flowers. But the collateral learning of the experience offers a far more profound reward: the ability to probe and telescope in difficult and ever-changing situations. It's not *what* the player is thinking about, but the *way* she's thinking.

At first glance, it might be tempting to connect the complexity of video games with the more familiar idea of "information overload" associated with the rise of electronic media. But a crucial difference exists. Information overload is a kind of backhanded compliment you'll often hear about today's culture: there's too much data flowing into our lives, but at least we're getting better at managing that data-stream, even if we may be approaching some kind of threshold point where our senses will simply be overwhelmed. This is a quantitative argument, not a qualitative one. It's nice to be able to watch TV, talk on the phone, and read your e-mail all at the same time, but it's a superficial skill, not a deep one. It usually involves skimming the surface of the incoming data, picking out the relevant details, and moving on to the next stream. Multimedia pioneer Linda Stone has coined a valuable term for this kind of processing: continuous partial attention. You're paying attention, but only partially. That lets you cast a wider net, but it also runs the risk of keeping you from really studying the fish.

Probing and telescoping represent another—equally important—tendency in the culture: the emergence of forms that encourage participatory thinking and analysis, forms that challenge the mind to make sense of an environment, not just play catch-up with the acceleration curve. I think for many people who do not have experience with them, games seem like an extension of the rapid-fire visual editing techniques pioneered by MTV twenty years ago: a seismic increase in images-per-second without a corresponding increase in analysis or sense-making. But the reality of MTV visuals is not that the eye learns to interpret all the images as they fly by, perceiving new relationships between them. Instead, the eye learns to tolerate chaos, to experience disorder as an aesthetic experience, the way the ear learned to appreciate distortion in music a generation before. To non-players, games bear a superficial resemblance to music videos: flashy graphics; the layered mix of image, music, and text; the occasional burst of speed, particularly during the pre-rendered opening sequences. But what you actually *do* in playing a game—the way your mind has to work—is radically different. It's not about tolerating or aestheticizing chaos; it's about finding order and meaning in the world, and making decisions that help create that order.

TELEVISION

THE INTERACTIVE NATURE of games means that they will inevitably require more decision-making than passive forms like television or film. But popular television shows—and to a slightly lesser extent, popular films—have also increased the cognitive work they demand from their audience, exercising the mind in ways that would have been unheard of thirty years ago. For someone loosely following the debate over the medium's cultural impact, the idea that television is actually improving our minds will sound like apostasy. You can't surf the Web or flip through a newsstand for more than a few minutes without encountering someone complaining about the surge in sex and violence on TV: from Tony Soprano to Janet Jackson. There's no questioning that the trend is real enough, though it is as old as television itself. In Newton Minow's famous "vast waste-land" speech from 1961, he described the content of current television programming as a "procession of . . . blood and thunder, mayhem, violence, sadism, murder"—this in the era of Andy Griffith, Perry Como, and Uncle Miltie. But evaluating the social merits of any medium and its pro-gramming can't be limited purely to questions of subject matter. There was nothing particularly redeeming in the subject matter of my dice baseball games, but they nonethe-less taught me how to think in powerful new ways. So if we're going to start tracking swear words and wardrobe malfunctions, we ought to at least include another line in the graph: one that charts the cognitive demands that televised narratives place on their viewers. That line, too, is trending upward at a dramatic rate.

Television may be more passive than video games, but there are degrees of passivity. Some narratives force you to do work to make sense of them, while others just let you settle into the couch and zone out. Part of that cognitive work comes from following multiple threads, keeping often densely interwoven plotlines distinct in your head as you watch. But another part involves the viewer's "filling in": making sense of information that has been either de-liberately withheld or deliberately left obscure. Narratives that require that their viewers fill in crucial elements take that complexity to a more demanding level. To follow the narrative, you aren't just asked to remember. You're asked to analyze. This is the difference between intelligent shows, and shows that force you to be intelligent. With many television classics that we associate with "quality" entertainment—*Mary Tyler Moore, Murphy Brown, Frasier*—the intelligence arrives fully formed in the words and actions of the characters onscreen. They say witty things to each other, and avoid lapsing into tired sitcom clichés, and we smile along in our living room, enjoying the company of these smart people. But assuming we're bright enough to understand the sentences they're saying—few of which are rocket science, mind you, or any kind of science, for that matter—there's no intellectual labor involved in enjoying the show as a viewer. There's no filling in, because the intellectual achievement exists entirely on the other side of the screen. You no more challenge your mind by watching these intelligent shows than you challenge your body watching *Monday Night Football*. The intellectual work is happening onscreen, not off.

But another kind of televised intelligence is on the rise. Recall the cognitive benefits conventionally ascribed to reading: attention, patience, retention, the parsing of narrative threads. Over the last half century of television's dominance over mass culture, programming on TV has steadily increased the demands it places on precisely these mental faculties. The nature of the medium is such that television will never improve its viewers' skills at translating letters into meaning, and it may not activate the imagination in the same way that a purely textual form does. But for all the other modes of mental exercise associated with reading, television is growing increasingly rigorous. And the pace is accelerating—thanks to changes in the economics of the television business, and to changes in the technology we rely on to watch.

This progressive trend alone would probably surprise someone who only read popular accounts of TV without watching any of it. But perhaps the most surprising thing is this: that the shows that have made the most demands on their audience have also turned out to be among the most lucrative in television history.

*   *   *

PUT ASIDE for a moment the question of why the marketplace is rewarding complexity, and focus first on the question of what this complexity looks like. It involves three primary elements: multiple threading, flashing arrows, and social networks.

Multiple threading is the most acclaimed structural convention of modern television programming, which is ironic because it's also the convention with the most debased pedigree. According to television lore, the age of multiple threads began with the arrival of *Hill Street Blues* in 1981, the Steven Bochco–created police drama invariably praised for its "gritty realism." Watch an episode of *Hill Street Blues* side by side with any major drama from the preceding decades—*Starsky and Hutch,* for instance, or *Dragnet*—and the structural transformation will jump out at you. The earlier shows follow one or two lead characters, adhere to a single dominant plot, and reach a decisive conclusion at the end of the episode. Draw an outline of the narrative threads in almost every *Dragnet* episode and it will be a single line: from the initial crime scene, through the investigation, to the eventual cracking of the case. A typical *Starsky and Hutch* episode offers only the slightest variation on this linear formula: the introduction of a comic subplot that usually appears only at the tail ends of the episode, creating a structure that looks like the graph below. The vertical axis represents the number of individual threads, and the horizontal axis is time.

STARSKY AND HUTCH (ANY EPISODE)

Starsky and Hutch includes a few other twists: While both shows focus almost exclusively on a single narrative, *Dragnet* tells the story entirely from the perspective of the investigators. *Starsky and Hutch,* on the other hand, oscillates between the perspectives of the cops and that of the criminals. And while both shows adhere religiously to the principle of narrative self-containment—the plots begin and end in a single episode—*Dragnet* takes the principle to a further extreme, introducing the setting and main characters with Joe Friday's famous voice-over in every episode.

A *Hill Street Blues* episode complicates the picture in a number of profound ways. The narrative weaves together a collection of distinct strands—sometimes as many as ten, though at least half of the threads involve only a few quick scenes scattered through the episode. The number of primary characters—and not just bit parts—swells dramatically. And the episode has fuzzy borders: picking up one or two threads from previous episodes at the outset, and leaving one or two threads open at the end. Charted graphically, an average episode looks like this:

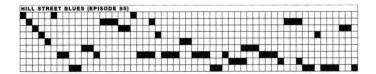

HILL STREET BLUES (EPISODE 85)

Critics generally cite *Hill Street Blues* as the origin point of "serious drama" native to the television medium—differentiating the series from the single episode dramatic programs from the fifties, which were Broadway plays performed in front of a camera. But the *Hill Street* innovations weren't all that original; they'd long played a defining role in popular television—just not during the evening hours. The structure of a *Hill Street* episode—and indeed all of the critically acclaimed dramas that followed, from *thirtysomething* to *Six Feet Under*—is the structure of a soap opera. *Hill Street Blues* might have sparked a new golden age of television drama during its seven-year run, but it did so by using a few crucial tricks that *Guiding Light* and *General Hospital* had mastered long before.

Bochco's genius with *Hill Street* was to marry complex narrative structure with complex subject matter. *Dallas* had already shown that the extended, interwoven threads of the soap opera genre could survive the weeklong interruptions of a prime-time show, but the actual content of *Dallas* was fluff. (The most probing issue it addressed was the now folkloric question of who shot JR.) *All in the Family* and *Rhoda* showed that you could tackle complex social issues, but they did their tackling in the comfort of the sitcom living room structure. *Hill Street* had richly drawn characters confronting difficult social issues, and a narrative structure to match.

Since *Hill Street* appeared, the multithreaded drama has become the most widespread fictional genre on prime time: *St. Elsewhere, thirtysomething, L.A. Law, Twin Peaks, NYPD Blue, ER, The West Wing, Alias, The Sopranos, Lost, Desperate Housewives.* The only prominent holdouts in drama are shows like *Law & Order* that have essentially updated the venerable *Dragnet* format, and thus remained anchored to a single narrative line. Since the early eighties, there has been a noticeable increase in narrative complexity in these dramas. The most ambitious show on TV to date— *The Sopranos*—routinely follows a dozen distinct threads over the course of an episode, with more than twenty recurring characters. An episode from late in the first season looks like this:

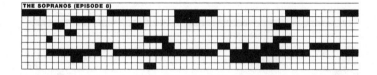

The total number of active threads equals the number of multiple plots of *Hill Street*, but here each thread is more substantial. The show doesn't offer a clear distinction between dominant and minor plots; each storyline carries its weight in the mix. The episode also displays a chordal mode of storytelling entirely absent from *Hill Street*: a single scene in *The Sopranos* will often connect to three different threads at the same time, layering one plot atop another. And every single thread in this *Sopranos* episode builds on events from previous episodes, and continues on through the rest of the season and beyond. Almost every sequence in the show connects to information that exists outside the frame of the current episode. For a show that spends as much time as it does on the analyst's couch, *The Sopranos* doesn't waste a lot of energy with closure.

Put these four charts together and you have a portrait of the Sleeper Curve rising over the past thirty years of popular television.

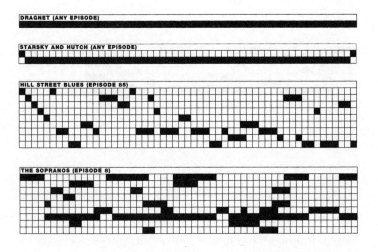

In a sense, this is as much a map of cognitive changes in the popular mind as it is a map of onscreen developments, as though the media titans had decided to condition our brains to follow ever larger numbers of simultaneous threads. Before *Hill Street,* the conventional wisdom among television execs was that audiences wouldn't be comfortable following more than three plots in a single episode, and indeed, the first test screening of the *Hill Street* pilot in May 1980 brought complaints from the viewers that the show was too complicated. Fast forward twenty years and shows like *The Sopranos* engage their audiences with narratives that make *Hill Street* look like *Three's Company*. Audiences happily embrace that complexity because they've been trained by two decades of multithreaded dramas.

Is there something apples-to-oranges in comparing a boutique HBO program like *The Sopranos* to a network primetime show like *Hill Street Blues*? Isn't the increase in complexity merely a reflection of the later show's smaller and more elite audience? I think the answer is no, for several reasons. First, measured by pure audience share, *The Sopranos* is a genuine national hit, regularly outdrawing network television shows in the same slot. Second, *Hill Street Blues* was itself a boutique show—the first step in NBC's immensely successful attempt in the early eighties to target an upscale demographic instead of the widest possible audience. The show was a cultural and critical success, but it spent most of its life languishing in the mid-thirties in the Nielsen TV ratings—and in its first season, the series finished eighty-third out of ninety-seven total shows on television. The total number of viewers for a *Sopranos* episode is not that different from that of an average episode of *Hill Street Blues*, even though the former's narrative complexity is at least twice that of the latter. (*The Sopranos* is even more complex on other scales, to which we will turn shortly.)

You can also measure the public's willingness to tolerate more complicated narratives in the success of shows such as *ER* or *24*. In terms of multiple threading, both shows usually follow around ten threads per episode, roughly comparable to *Hill Street Blues*. But *ER* and *24* are bona fide hits, regularly appearing in the Nielsen top twenty. In 1981, you could weave together three major narratives and a half dozen supporting plots over the course of an hour on prime time, and cobble together enough of an audience to keep the show safe from cancellation. Today you can challenge the audience to follow a more complicated mix, and build a juggernaut in the process.

Multithreading is the most celebrated structural feature of the modern television drama, and it certainly deserves some of the honor that has been doled out to it. When we watch TV, we intuitively track narrative-threads-per-episode as a measure of a given show's complexity. And all the evidence suggests that this standard has been rising steadily over the past two decades. But multithreading is only part of the story.

*   *   *

A FEW YEARS after the arrival of the first-generation slasher movies—*Halloween, Friday the 13th*—Paramount

released a mock-slasher flick, *Student Bodies,* which paro-
died the genre just as the *Scream* series would do fifteen
years later. In one scene, the obligatory nubile teenage
babysitter hears a noise outside a suburban house; she opens
the door to investigate, finds nothing, and then goes back inside. As the door shuts behind her, the camera swoops in on
the doorknob, and we see that she's left the door unlocked.
The camera pulls back, and then swoops down again, for
emphasis. And then a flashing arrow appears on the screen,
with text that helpfully explains: "Door Unlocked!"

That flashing arrow is parody, of course, but it's merely
an exaggerated version of a device popular stories use all the
time. It's a kind of narrative signpost, planted conveniently
to help the audience keep track of what's going on. When
the villain first appears in a movie emerging from the shad-
ows with ominous, atonal music playing—that's a flashing
arrow that says: "bad guy." When a sci-fi script inserts a
non-scientist into some advanced lab who keeps asking the
science geeks to explain what they're doing with that particle accelerator—that's a flashing arrow that gives the audience precisely the information they need to know in order
to make sense of the ensuing plot. ("Whatever you do, don't
spill water on it, or you'll set off a massive explosion!")
Genre conventions function as flashing arrows; the *Student
Bodies* parody works because the "door unlocked" text is
absurd overkill—we've already internalized the rules of the
slasher genre enough to know that nubile-babysitter-in-

suburban-house inevitably leads to unwanted visitors. Heist movies traditionally deliver a full walk-through of the future crime scene, complete with architectural diagrams, so you'll know what's happening when the criminals actually go in for the goods.

These hints serve as a kind of narrative handholding. Implicitly, they say to the audience, "We realize you have no idea what a particle accelerator is, but here's the deal: all you need to know is that it's a big fancy thing that explodes when wet." They focus the mind on relevant details: "Don't worry about whether the babysitter is going to break up with her boyfriend. Worry about that guy lurking in the bushes." They reduce the amount of analytic work you need to make sense of a story. All you have to do is follow the arrows.

By this standard, popular television has never been harder to follow. If narrative threads have experienced a population explosion over the past twenty years, flashing arrows have grown correspondingly scarce. Watching our pinnacle of early eighties TV drama, *Hill Street Blues,* there's an informational *wholeness* to each scene that differs markedly from what you see on shows like *The West Wing* or *The Sopranos* or *Alias* or *ER. Hill Street* gives you multiple stories to follow, as we've seen, but each event in those stories has a clarity to it that is often lacking in the later shows.

This is a subtle distinction, but an important one, a facet of the storyteller's art that we sometimes only soak up unconsciously. *Hill Street* has ambiguities about future events:

Will the convicted serial killer be executed? Will Furillo marry Joyce Davenport? Will Renko catch the health inspector who has been taking bribes? But the present tense of each scene explains itself to the viewer with little ambiguity. You may not know the coming fate of the health inspector, but you know why Renko is dressing up as a busboy in the current scene, or why he's eavesdropping on a kitchen conversation in the next. There's an open question or a mystery driving each of these stories—how will it all turn out?—but there's no mystery about the immediate activity on the screen.

A contemporary drama like *The West Wing*, on the other hand, constantly embeds mysteries into the present-tense events: you see characters performing actions or discussing events about which crucial information has been deliberately withheld. Appropriately enough, the extended opening sequence of the *West Wing* pilot revolved around precisely this technique: you're introduced to all the major characters (Toby, Josh, CJ) away from the office, as they each receive the enigmatic message that "POTUS has fallen from a bicycle." *West Wing* creator Aaron Sorkin—who amazingly managed to write every single episode through season four—deliberately withholds the information that all these people work at the White House, and that POTUS stands for "President of the United States," until the very last second before the opening credits run. Granted, a viewer tuning in to a show called *The West Wing* probably suspected that there was going to be some kind of White House connection, and a few political aficionados might have already been familiar with the acronym POTUS. But that opening sequence established a structure that Sorkin used in every subsequent episode, usually decorated with deliberately opaque information. The open question posed by these sequences is not: How will this turn out in the end? The question is: What's happening right now?

In practice, the viewers of shows like *Hill Street Blues* in the eighties no doubt had moments of confusion where the sheer number of simultaneous plots created present-tense mystery: we'd forget why Renko was wearing that busboy outfit because we'd forgotten about the earlier sequence introducing the undercover plot. But in that case, the missing information got lost somewhere between our perceptual systems and our short-term data storage. The show gave us a clear vista on to the narrative events; if that view fogged over, we had only our memory to blame. Sorkin's shows, on the other hand, are the narrative equivalent of fog machines. You're *supposed* to be in the dark. Anyone who has watched more than a handful of *West Wing* episodes closely will know the feeling: scene after scene refers to some clearly crucial piece of information—the cast members will ask each other if they saw "the interview" last night, or they'll make enigmatic allusions to the McCarver case—and after the sixth reference, you'll find yourself wishing you could rewind the tape to figure out what they're talking about, assuming you've missed something. And then you realize that you're supposed to be confused.

The clarity of *Hill Street* comes from the show's subtle integration of flashing arrows, while *West Wing*'s murkiness comes from Sorkin's cunning refusal to supply them. The roll call sequence that began every *Hill Street* episode is most famous for the catchphrase "Hey, let's be careful out there." But that opening address from Sergeant Esterhaus (and in later seasons, Sergeant Jablonski) performed a crucial function, introducing some of the primary threads and providing helpful contextual explanations for them. Critics at the time remarked on the disorienting, documentary-style handheld camerawork used in the opening sequence, but the roll call was ultimately a comforting device for the show, training wheels for the new complexity of multithreading.

Viewers of *The West Wing* or *Lost* or *The Sopranos* no longer require those training wheels, because twenty-five years of increasingly complex television has honed their analytic skills. Like those video games that force you to learn the rules while playing, part of the pleasure in these modern television narratives comes from the cognitive labor you're forced to do filling in the details. If the writers suddenly dropped a hoard of flashing arrows onto the set, the show would seem plodding and simplistic. The extra information would take the fun out of watching.

This deliberate lack of handholding extends down to the micro level of dialogue as well. Popular entertainment that addresses technical issues—whether they are the intricacies of passing legislation, or performing a heart bypass, or operating a particle accelerator—conventionally switches between two modes of information in dialogue: texture and substance. Texture is all the arcane verbiage provided to convince the viewer that they're watching Actual Doctors At Work; substance is the material planted amid the background texture that the viewer needs to make sense of the plot.

Ironically, the role of texture is sometimes to be directly *irrelevant* to the concerns of the underlying narrative, the more irrelevant the better. Roland Barthes wrote a short essay in the sixties that discussed a literary device he called the "reality effect," citing a description of a barometer from Flaubert's short story "A Simple Heart." In Barthes's description, reality effects are designed to create the aura of real life through their sheer meaninglessness: the barometer doesn't play a role in the narrative, and it doesn't symbolize anything. It's just there for background texture, to create the illusion of a world cluttered with objects that have no narrative or symbolic meaning. The technical banter that proliferates on shows like *The West Wing* or *ER* has a comparable function; you don't need to know what it means when the surgeons start shouting about OPCAB and saphenous veins as they perform a bypass on *ER*; the arcana is there to create the illusion that you are watching real doctors. For these shows to be enjoyable, viewers have to be comfortable knowing that this is information they're not supposed to understand.

Conventionally, narratives demarcate the line between texture and substance by inserting cues that flag or translate the important data. There's an unintentionally comical moment in the 2004 blockbuster *The Day After Tomorrow* where the beleaguered climatologist (played by Dennis Quaid) announces his theory about the imminent arrival of a new ice age to a gathering of government officials. His oration ends with the line: "We may have hit a critical desalinization threshold!" It's the kind of thing that a climatologist might plausibly say—were he dropped into an alternative universe where implausible things like instant ice ages actually happened—but for most members of the audience, the phrase "critical desalinization threshold" is more likely to elicit a blank stare than a spine tingle. And so the writer/director Roland Emmerich—a master of brazen arrow-flashing—has a sidekick official next to Quaid follow with the obliging remark: "That would explain all the extreme weather we're having." They might as well have had a flashing "Door Unlocked!" arrow on the screen.

The dialogue on shows like *The West Wing* and *ER,* on the other hand, doesn't talk down to its audience. It rushes by, the words accelerating in sync with the high-speed tracking shots that glide through the corridors and operating rooms. The characters talk faster in these shows, but the truly remarkable thing about the dialogue is not purely a matter of speed; it's the willingness to immerse the audience in information that most viewers won't understand. Here's a typical scene from *ER*:

> Cut to KERRY bringing in a young girl, CARTER and LUCY run up.
>
> The girl's parents are also present.
>
> KERRY: Sixteen-year-old unconcious, history of villiari treesure.
>
> CARTER: Glucyna coma?
>
> KERRY: Looks like it.
>
> MR. MAKOMI: She was doing fine until six months ago.
>
> CARTER: What medication is she on?
>
> MRS. MAKOMI: Emphrasylim, tobramysim, vitamins A, D, and K.
>
> LUCY: The skin's jaundiced.
>
> KERRY: Same with sclera, does her breath smell sweet?
>
> CARTER: Peder permadicis?
>
> KERRY: Yeah.
>
> LUCY: What's that?
>
> KERRY: Liver's shut down, let's dip her urine. (To CARTER) It's getting a little crowded in here, why don't you deal with the parents, please. Set lactolose, 30 ccs per mg.
>
> CARTER: We're gonna give her some medicine to clean her blood, why don't you come with me?

CARTER leads the MAKOMIs out of the trauma room,
  LUCY also follows him

KERRY: Blood doesn't seem to clot.

MR. MAKOMI: She's bleeding inside?

CARTER: The liver failure is causing her blood not
  to clot.

MRS. MAKOMI: Oh God.

CARTER: Is she on the transplant list?

MR. MAKOMI: She's been status 2a for six months but
  they haven't been able to find her a match.

CARTER: Why not, what's her blood type?

MR. MAKOMI: AB.

CARTER and LUCY stare at each other in disbelief.

Cut to MARK working on a sleeping patient. AMANDA
  walks in.

There are flashing arrows here, of course—"The liver
failure is causing her blood not to clot"—but the ratio of
medical jargon to layperson translation is remarkably high,
and as in so many of these narratives, you don't figure out
what's really happening until the second half of the scene.
There's a kind of implicit trust formed between the show
and its viewers, a tolerance for planned ambiguity. That tolerance takes work: you need to be able to make assessments
on the fly about the role of each line, putting it in the "substance" or "texture" slot. You have to know what you're not

supposed to know. If viewers weren't able to make those assessments in real time, *ER* would be an unbearable mess; you'd have to sit down every Thursday night with a medical dictionary at hand. ("Is *peder permadicis* spelled with a *d* or a *t?*")

From a purely narrative point of view, the decisive line in that scene arrives at the very end: "AB." The sixteen-year-old's blood type connects her to an earlier plotline, involving a cerebral hemorrhage victim who—after being dramatically revived in one of the opening scenes—ends up brain dead. Fifteen minutes before the liver-failure scene above, Doug and Carter briefly discuss harvesting the hemorrhage victim's organs for transplants, and make a passing reference to his blood type being the rare AB. (Thus making him an unlikely donor.) The twist here revolves around a statistically unlikely event happening at the ER—an otherwise perfect liver donor showing up just in time to donate his liver to a recipient with the same rare blood type. But the show reveals this twist with a remarkable subtlety. To make sense of that last "AB" line—and the look of disbelief on Carter's and Lucy's faces—you have to recall a passing remark uttered fifteen minutes before regarding a character who belongs to a completely different thread.

It would have been easy enough to insert an explanatory line at the end of the scene: "That's the same blood type as our hemorrhage victim!" And in fact, had *ER* been made twenty or thirty years ago, I suspect the writers would have added precisely such a line. But that kind of crude subtitling would go against the narrative ethos of shows like *ER*. In these modern narratives, part of the pleasure comes from the audience's "filling in." These shows may have more blood and guts than popular TV had a generation ago, and some of the sexual content today would have been inappropriate in a movie theater back then—much less on prime-time TV. But when it comes to storytelling, these shows possess a quality that can only be described as subtlety and discretion.

It's not a headline you often see—"Pop TV More Subtle and Discreet Than Ever Before!"—but ignoring these properties means overlooking one of the most vital developments in modern popular narrative. You'll sometimes hear people refer fondly to the "simpler" era of television's alleged heyday, the days of *Dragnet* and *I Love Lucy*. They mean "simpler" in an ethical sense: there were no sympathetic mob bosses on *Dragnet*, no custody battles on *Lucy*. But when you watch these shows next to today's television, the other sense of "simpler" applies as well: they require less mental labor to make sense of what's going on. Watch *Starsky and Hutch* or *Dragnet* after watching *The Sopranos* and you'll feel as though you're being condescended to— because the creators of those shows are imagining an "ideal viewer" who has not benefited from decades of the Sleeper Curve at work. They kept it simple because they assumed their audience at the time wasn't ready for anything more complicated.

In this, they were probably right.

*   *   *

TELEVISION DRAMA is the most dramatic instance of the Sleeper Curve, but you can see a comparable shift toward increased complexity in most of the sitcoms that have flourished over the past decade. Compare the way comedy unfolds in recent classics like *Seinfeld* and *The Simpsons*—along with newer critics' faves like *Scrubs* or *Arrested Development*—to earlier sitcoms like *All in the Family* or *Mary Tyler Moore*. The most telling way to measure these shows' complexity is to consider how much external information the viewer must draw upon to "get" the jokes in their entirety. Anyone can sit down in front of most run-of-the-mill sitcoms—*Home Improvement,* say, or *Three's Company*—and the humor will be immediately intelligible, since it consists mostly of characters being sarcastic to each other. The jokes themselves make no reference to anything outside the frame of the conversation that contains them—beyond the bare-bones "situation" that the sitcom itself is grounded in. (A guy pretends that he's gay so he can shack up with two women.) To parse the humor of more nuanced shows—*Cheers* or *Friends,* for example—the scripts will sometimes demand that you know some basic biographical information about the characters. (Carla will make a snotty reference to Sam Malone's sobriety, without bothering to explain to the audience that he once had a drinking problem; or Rachel will allude to Monica's overweight childhood.) Nearly every extended sequence in *Seinfeld* or *The Simpsons,* however, will contain a joke that makes sense only if the viewer fills in the proper supplementary information—information that is deliberately withheld from the viewer. If you haven't seen the "Mulva" episode, or if the name "Art Vandelay" means nothing to you, then the subsequent references—many of them arriving years after their original appearance—will pass on by unappreciated.

At first glance, this looks like the soap opera tradition of plotlines extending past the frame of individual episodes, but in practice the device has a different effect. Knowing that George uses the alias Art Vandelay in awkward social situations doesn't help you understand the plot of the current episode; you don't draw on past narratives to understand the events of the present one. In the 180 *Seinfeld* episodes that aired, seven contain references to Art Vandelay: in George's actually referring to himself with that alias or invoking the name as part of some elaborate lie. He tells a potential employer at a publishing house that he likes to read the fiction of Art Vandelay, author of *Venetian Blinds*; in another, he tells an unemployment insurance caseworker that he's applied for a latex salesman job at Vandelay Industries. For storytelling purposes, the only thing that you need to know here is that George is lying in a formal interview; any fictitious author or latex manufacturer would suffice. But the *joke* arrives through the echo of all those earlier Vandelay references; it's funny because it's making a subtle nod to past events held offscreen. It's what we'd call in a real-world context an "in-joke"—a joke that's funny only to people who get the reference. And in this case, the reference is to a few fleeting lines in a handful of episodes—most of which aired years before. Television comedy once worked on the scale of thirty seconds: you'd have a setup line, and then a punch line, and then the process would start all over again. With *Seinfeld,* the gap between setup and punch line could sometimes last five years.

These layered jokes often point beyond the bounds of the series itself. According to one fan site that has exhaustively chronicled these matters, the average *Simpsons* episode includes around eight gags that explicitly refer to movies: a plotline, a snippet of dialogue, a visual pun on a famous cinematic sequence (*Seinfeld* featured a number of episodes that mirrored movie plots, including *Midnight Cowboy* and *JFK*). The Halloween episodes have historically been the most baroque in their cinematic allusions, with the all-time champ being an episode from the 1995 season, integrating material from *Attack of the 50 Foot Woman, Godzilla, Ghostbusters, Nightmare on Elm Street, The Pagemaster, Maximum Overdrive, The Terminator* and *Terminator 2,*

Alien III, Tron, Beyond the Mind's Eye, The Black Hole, Poltergeist, Howard the Duck, and *The Shining.*

The film parodies and cultural sampling of *The Simpsons* usually get filed away as textbook postmodernism: media riffing on other media. But the Art Vandelay jokes from *Seinfeld* don't quite fit the same postmodern mold: they aren't references that jump from one fictional world to another; they're references that jump back in time within a single fictional world. I think it's more instructive to see both these devices as sharing a key attribute: they are comic devices that reward further scrutiny. The show gets funnier the more you study it—precisely because the jokes point outside the immediate context of the episode, and because the creators refuse to supply flashing arrows to translate the gags for the uninitiated. Earlier sitcoms merely demanded that you kept the basic terms of the situation clear on your end; beyond that information you could be an amnesiac and you weren't likely to miss anything. Shows like *Seinfeld* and *The Simpsons* offered a more challenging premise to their viewers: You'll enjoy this more if you're capable of remembering a throwaway line from an episode that aired three years ago, or if you notice that we've framed this one scene so that it echoes the end of *Double Indemnity.* The jokes come in layers: you can watch that 1995 Halloween episode and miss all the film riffs and still enjoy the show, but it's a richer, more rewarding experience if you're picking them up.

That layering enabled *Seinfeld* and *The Simpsons* to retain both a broad appeal and the edgy allure of cult classics. The mainstream audiences chuckle along to that wacky Kramer, while the diehard fans nudge-and-wink at each Superman aside. But that complexity has another, equally important, side effect: the episodes often grow *more* entertaining on a second or third viewing, and they can still reveal new subtleties on the fifth or sixth. The subtle intertwinings of the plots seem more nimble if you know in advance where they're headed, and the more experience you have with the series as a whole, the more likely you are to catch all the insider references.

In November 1997, NBC aired an episode of *Seinfeld* called "The Betrayal," in which the scenes were presented in reverse chronological order. If the *Seinfeld* formula often involved setups followed by punch lines that arrived years later, "The Betrayal" took a more radical approach: punch lines that arrived *before* their setups. You'd see Kramer begging Newman to protect him from a character called "FDR," and only find out why ten minutes later, when you're shown an "earlier" scene where FDR gives Kramer the evil eye at a birthday party. The title of the episode (and the name of one of the characters) was a not-so-subtle nod to the Harold Pinter play *Betrayal*, which told the story of a love triangle as a reverse chronology. But comedies are different from dramas in their relationship to time: a dramatic event with no context is a mystery—the withheld information can heighten the drama. But a punch line with no context is not a joke. Nearly unwatchable the first time around, "The Betrayal" became coherent only on a second viewing—and it took three solid passes before the jokes started to work. You'd see the punch line delivered onscreen, and you'd fill in the details of the setup on your own.

"The Betrayal" was a watershed in television programming, assembling all the elements of modern TV complexity in one thirty-minute sitcom. The narrative wove together seven distinct threads, withheld crucial information in almost every sequence, and planted jokes that had multiple layers of meaning. As the title implied, these were storytelling devices that you would have found only in avant-garde narrative thirty or forty years ago: in Pinter, or Alain Robbe-Grillet, or Godard. You might have been able to fill a small theater in Greenwich Village with an audience willing to parse all that complexity in 1960, but only if the *Times* had given the play a good review that week. Forty years later, NBC puts the same twisted narrative structure on prime-time television, and 15 million people lap it up.

A few popular sitcoms have done well with the traditional living room banter of yesteryear: *Everybody Loves Raymond* comes to mind. But most comedies that have managed to achieve both critical and commercial success—*Scrubs, The Office, South Park, Will & Grace, Curb Your Enthusiasm*—have almost without exception taken their structural cues from *The Simpsons* instead of *Three's Com-*

pany: creating humor with a half-life longer than fifteen seconds, drawing on intricate plotlines and obscure references. But the sitcom genre as a whole has wilted in the past few years, as television execs turned their focus to the new—and oft-abused—ratings champ: reality programming.

*   *   *

SKEPTICS MIGHT ARGUE that I have stacked the deck here by focusing on relatively highbrow titles like *The Simpsons* or *The West Wing,* when in fact the most significant change in the last five years of narrative entertainment has nothing to do with complex dramas or self-referential sitcoms. Does the contemporary pop cultural landscape look quite as promising if the representative TV show is *Joe Millionaire* instead of *The West Wing*?

I think it does, but to answer that question properly, you have to avoid the tendency to sentimentalize the past. When people talk about the golden age of television in the early seventies—invoking shows like *Mary Tyler Moore* and *All in the Family*—they forget to mention how awful most television programming was during much of that decade. If you're going to look at pop culture trends, you have to compare apples to apples, or in this case, lemons to lemons. If *Joe Millionaire* is a dreadful show that has nonetheless snookered a mass audience into watching it, then you have to compare it to shows of comparable quality and audience reach from thirty years ago for the trends to be meaningful. The relevant comparison is not between *Joe Millionaire* and *M*A*S*H*; it's between *Joe Millionaire* and *The Price Is Right,* or between *Survivor* and *The Love Boat.*

What you see when you make these head-to-head comparisons is that a rising tide of complexity has been lifting programming both at the bottom of the quality spectrum and at the top. *The Sopranos* is several times more demanding of its audiences than *Hill Street* was, and *Joe Millionaire* has made comparable advances over *Battle of the Network Stars.* This is the ultimate test of the Sleeper Curve theory: even the crap has improved.

How might those improvements be measured? To take stock of this emerging genre, once again we have to paint our portrait of the rhinoceros carefully, to capture why people really get hooked on these shows. Because I think the appeal is often misunderstood. The conventional wisdom is that audiences flock to reality programming because they enjoy the prurient sight of other people being humiliated on national TV. This indeed may be true for gross-out shows like *Fear Factor,* where contestants lock themselves into vaults with spiders or consume rancid food for their fifteen minutes of fame. But for the most successful reality shows—*Survivor* or *The Apprentice*—the appeal is more sophisticated. That sophistication has been difficult to see, because reality programming, too, has suffered from our tendency to see emerging genres as "pseudo" versions of earlier genres, as McLuhan diagnosed. When reality programming first burst on the scene, it was traditionally compared with the antecedent form of the documentary film. Naturally, when you compare *Survivor* with *Shoah*, *Survivor* comes up short. But reality shows do not represent reality the way documentaries represent reality. *Survivor*'s relationship to reality is much closer to the relationship between professional sports and reality: highly contrived, rule-governed environments where (mostly) unscripted events play out.

Thinking of reality shows in the context of games gives us useful insight into the merits of the genre, as opposed to the false comparisons to Barbara Koppel films and *Capturing the Friedmans*. Perhaps the most important thing that should be said about reality programming is that the format is reliably structured like a video game. Reality television provides the ultimate testimony to the cultural dominance of games in this moment of pop culture history. Early television took its cues from the stage: three-act dramas, or vaudeville-like acts with rotating skits and musical numbers. In the Nintendo age, we expect our televised entertainment to take a new form: a series of competitive tests, growing more challenging over time. Many reality shows borrow a subtler device from gaming culture as well: the rules aren't fully established at the outset. You learn as you play. On a show like *Survivor* or *The Apprentice*, the participants—and the audience—know the general objective of the series, but each episode involves new challenges that haven't been ordained in advance. The final round of season one of *The Apprentice,* for instance, threw a monkeywrench into the strategy that had governed the play up until that point, when Trump announced that the two remaining apprentices would have to assemble and manage a team of subordinates who had already been fired in earlier episodes of the show. All of a sudden the overarching objective of the game—do anything to avoid being fired—presented a potential conflict to the remaining two contenders: the structure of the final round favored the survivor who had maintained the best relationships with his comrades. Suddenly, it wasn't enough just to have clawed your way to the top; you had to have made friends while clawing.

The rules and conventions of the reality genre are in flux, and that unpredictability is part of the allure. This is one way in which reality shows differ dramatically from their game show ancestors. When new contestants walked onstage for *The Price Is Right* or *Wheel of Fortune,* no ambiguity existed about the rules of engagement; everyone knew how the game was played—the only open question was who would be the winner, and what fabulous prizes they'd take home. In reality TV, the revealing of the game's rules is part of the drama, a deliberate ambiguity that is celebrated and embraced by the audience. The original *Joe Millionaire* put a fiendish spin on this by undermining the most fundamental convention of all—that the show's creators don't openly lie to the contestants about the prizes—by inducing a construction worker to pose as a man of means while fifteen women competed for his attention.

Reality programming borrowed another key ingredient from games: the intellectual labor of probing the system's rules for weak spots and opportunities. As each show discloses its conventions, and each participant reveals his or her personality traits and background, the intrigue in watching comes from figuring out how the participants should best navigate the environment that's been created for them. The pleasure in these shows comes not from watching other human beings humiliated on national television; it comes from depositing other human beings in a complex, high-stakes environment where no established strategies exist, and watching them find their bearings. That's why the water-cooler conversation about these shows invariably tracks in on the strategy displayed on the previous night's episode: Why did Kwame pick Omarosa in that final round? What devious strategy is Richard Hatch concocting now?

Some of that challenge comes from an ever-changing system of rules, but it also comes from the rich social geography that all reality programming explores. In this one respect, the reality shows exceed the cognitive demands of the video games, because the games invariably whittle away at the branches of social contact. In the gameworld, you're dealing with real people through the mediating channels of 3D graphics and text chat; reality shows drop flesh-and-blood people into the same shared space for months at a time, often limiting their contact with the outside world. Reality program participants are forced to engage face-to-face with their comrades, and that engagement invariably taps their social intelligence in ways that video games can only dream of. And that social chess becomes part of the audience's experience as well. This, of course, was the appeal of that pioneering reality show, MTV's *The Real World,* which didn't need contests and fabulous prizes to lure its viewers; it just needed a group of people thrust together in a new space and forced to interact with one another.

The role of audience participation is one of those properties that often ends up neglected when the critics assess these shows. If you take reality programming to be one long extended exercise in public humiliation, then the internal monologue of most viewers would sound something like this: "Look at this poor fool—what a jackass!" Instead, I suspect those inner monologues are more likely to project the viewer into the show's world; they're participatory, if only hypothetically so: "If I were choosing who to kick off the island, I'd have to go with Richard." You assess the social geography and the current state of the rules, and you imagine how you would have played it, had you made it through the casting call. The pleasure and attraction of that kind of involvement differ from the narrative pleasure of the sitcom: the appeal of *Happy Days* doesn't come from imagining how you might have improved on the pep talk that Fonzie gives Richie over lunch at Al's. But in the world of reality programming, that projection is a defining part of the audience's engagement with the show.

Old-style game show viewers also like to imagine themselves as participants; people have been shouting out the answers in their living rooms since the days of *21*. (Reality programming embraces and extends the logic of game shows, just as shows like *The Sopranos* and *Six Feet Under* expand on the template originally created by the soap opera.) But the rules and the "right answers" have increased in complexity since Herbert Stempel took his famous dive. "Playing" a reality show requires you to both adapt to an ever-changing rulebook, and scheme your way through a minefield of personal relationships. To succeed in a show like *The Apprentice* or *Survivor,* you need social intelligence, not just a mastery of trivia. When we watch these shows, the part of our brain that monitors the emotional lives of the people around us—the part that tracks subtle shifts in intonation and gesture and facial expression—scrutinizes the action on the screen, looking for clues. We trust certain characters implicitly, and vote others off the island in a heartbeat. Traditional narrative shows also trigger emotional connections to the characters, but those connections don't have the same participatory effect, because traditional narratives aren't explicitly about *strategy*. The phrase "Monday-morning quarterbacking" was coined to describe the engaged feeling spectators have in relation to games as opposed to stories. We absorb stories, but we second-guess games. Reality programming has brought that second-guessing to prime time, only the game in question revolves around *social* dexterity rather than the physical kind.

Reality programming unfolds in the most artificial of environments: tropical islands swarming with invisible camera crews; castles populated by beautiful single women and one (fake) millionaire bachelor. But they nonetheless possess an emotional authenticity that is responsible for much of their appeal. At the peak moments—when Joe Millionaire reveals his true construction worker identity; when a contestant gets kicked off the island late in a *Survivor* series—the camera zooms in on the crestfallen face of the unlucky contestant, and what you see for a few fleeting seconds is something you almost never see in prime-time entertainment: a display of genuine emotion written on someone's face. The thrill of it is the thrill of something real and unplanned bursting out in the most staged and sterile of places, like a patch of wildflowers blooming in a parking lot. I find these moments cringe-inducing, because the emotions are so raw, but also bizarrely hypnotic: these are people who have spent the last six months dreaming of a life-changing event, only to find at the last minute that they've fallen short. The thrill of reality TV is seeing their face at the moment they get the news; the thrill of thinking, "This is actually happening." Next to that kind of emotional intensity, it's no wonder the sitcom—with its one-liners and canned laughter—has begun to wither.

I admit that there's something perverse in these moments, something like the frisson that pornography used to induce before it became a billion-dollar industry: what electrifies is the sense that *this is actually happening.* In a world of forgeries, this person on the screen isn't faking it, at least for that split second as the emotion washes over his face. You cover your eyes because the authenticity of the feeling is almost too hot for the medium.

"Split second" is the appropriate timescale here; the intelligence that the reality shows draw upon is the intelligence of microseconds: the revealing glance, the brief look of disbelief, a traitorous frown quickly wiped from a face. Humans express the full complexity of their emotions through the unspoken language of facial expressions, and we know from neuroscience that parsing that language—in all of its subtlety—is one of the great accomplishments of the human brain. One measure of this intelligence is called AQ, short for "autism quotient." People with low AQ scores are particularly talented at reading emotional cues, anticipating the inner thoughts and feelings of other people, a skill that is sometimes called mind reading. (Autistic people suffer from a diminished capacity for reading the language of facial expressions, which is why a high AQ score implies worse mind reading skills.) AQ can be seen as a subset of Daniel Goleman's concept of "emotional intelligence"; being smart is sometimes about doing complicated math in our heads, or making difficult logical decisions, but an equally important measure of practical intelligence is our ability to assess—and respond appropriately—to other people's emotional signals.

When you look at reality TV through the lens of AQ, the cognitive demands of the genre become much easier to appreciate. We had game shows to evaluate and reward our knowledge of trivia, and professional sports to reward our physical intelligence. Reality shows, in turn, challenge our emotional intelligence and our AQ. They are, in a sense, elaborately staged group psychology experiments, where at the end of the session the subjects get a million dollars and a week on the cover of *People* instead of a fifty-dollar stipend. The shows seem so fresh to today's audience because they tap this crucial faculty of the mind in ways that ordinary dramas or comedies rarely do—borrowing the participatory format of the game show while simultaneously challenging our emotional IQ. *The Apprentice* may not be the smartest show in the history of television, but it nonetheless forces you to think while you watch it, to work through the social logic of the universe it creates on the screen. And compared with *The Price Is Right* or *Webster,* it's an intellectual masterpiece.

Television turns out to be a brilliant medium for assessing other people's emotional intelligence or AQ—a property that is too often ignored when critics evaluate the medium's carrying capacity for thoughtful content. Part of this neglect stems from the age-old opposition between intelligence and emotion: intelligence is following a chess match or imparting a sophisticated rhetorical argument on a matter of public policy; emotions are the province of soap operas. But countless studies have demonstrated the pivotal role that emotional intelligence plays in seemingly high-minded arenas: business, law, politics. Any profession that involves regular interaction with other people will place a high premium on mind reading and emotional IQ. Of all the media available to us today, television is uniquely suited for conveying the fine gradients of these social skills. A book will give you a better vista of an individual's life story, and a newspaper op-ed is a better format for a rigorous argument, but if you're trying to evaluate a given person's emotional IQ and you don't have the option of sitting down with them in person, the tight focus of television is your best bet. Reality programming has simply recognized that intrinsic strength and built a whole genre around it.

Politics, too, has gravitated toward the television medium's emotional fluency. This is often derided as a coarsening or sentimentalizing of the political discourse, turning the rational debate over different political agendas into a Jerry Springer confessional. The days of the Lincoln–Douglas debates have given way to "Boxers or briefs?" The late Neil Postman described this sorry trend as the show-businessification of politics in his influential 1985 book, *Amusing Ourselves to Death*. In Postman's view, television is a medium of cosmetics, of surfaces, an endless replay of the Nixon–Kennedy debates, where the guy with the best makeup always wins. "Although the Constitution makes no mention of it, it would appear that fat people are now effectively excluded from running for high political office," he writes. "Probably bald people as well. Almost certainly those whose looks are not significantly enhanced by the cosmetician's art. Indeed, we may have reached the point where cosmetics has replaced ideology as the field of expertise over which a politician must have competent control."

No doubt some of what Postman says is true, though Bill Clinton did manage to eke out a successful political career while battling a minor weight problem. Television lets you see the physical characteristics of the people you're voting for with an accuracy unrivaled by any medium to date. To be sure, this means that physically repulsive individuals have suffered on election day. (Of course, it also means a commander in chief will no longer be able to conceal from the American people the simple fact that he can't *walk*.)

But the visibility of the medium extends beyond hairstyles and skin tone. When we see our politicians in the global living room of televised intimacy, we're able to detect more profound qualities in them: not just their grooming, but their emotional antennae—their ability to connect, outfox, condemn, or console. We see them as emotional mind readers, and there are few qualities in an individual more predictive of their ability to govern a country, because mind reading is so central to the art of persuasion. Presidents make formal appearances and sit for portraits and host galas, but their day-to-day job is motivating and persuading other people to follow their lead. To motivate and persuade you have to have an innate radar for other people's mental states. For an ordinary voter, it's almost impossible to get a sense for a given candidate's emotional radar without seeing them in person, in an unscripted setting. You can't get a sense of a candidate's mind reading skills by watching them give a memorized stump speech, or seeing their thirty-second ads, or God knows reading their campaign blog posts. But what *does* give you that kind of information is the one-on-one television interview format—*Meet the Press* and *Charlie Rose,* of course, but probably more effectively, *Oprah,* because the format is more social and free-flowing.

So what we're getting out of the much-maligned Oprahization of politics is not boxers-or-briefs personal trivia—it's crucial information about the emotional IQ of a potential president, information we had almost no access to until television came along and gave us that tight focus. Reading the transcript of the Lincoln–Douglas debates certainly conveyed the agility of both men's minds, and the ideological differences that separated them. But I suspect they conveyed almost no information about how either man would run a cabinet meeting, or what kind of loyalty they would inspire in their followers, or how they would resolve an internal dispute. Thirty minutes on a talk show, on the other hand, might well convey all that information—

because our brains are so adept at picking up those emotional cues. Physically unappealing candidates may not fare as well in this environment. (Lyndon Johnson would have a tough time of it today.) But the candidates who do pass the appearance test are judged by a higher, more discriminating standard—not just the color of their skin, but the content of their character.

That's not to imply that all political debate should be reduced to talk-show banter; there's still plenty of room for position papers and formal speeches. But we shouldn't underestimate the information conveyed by the close-ups of the unscripted television appearance. That first Nixon–Kennedy debate has long been cited as the founding moment of the triumph of image over substance—among all those TV viewers who thought Nixon's sweating and five-o'clock shadow made him look shifty and untrustworthy. But what if we've had it wrong about that debate? What if it wasn't Nixon's lack of makeup that troubled the TV watchers? After all, Nixon did turn out to be shifty and untrustworthy in the end. Perhaps all those voters who thought he had won after they heard the debate on the radio or read the transcript in the papers simply didn't have access to the range of emotional information conveyed by television. Nixon lost on TV because he didn't *look* like someone you would want as president, and where emotional IQ is concerned, looks don't always deceive.

❖   ❖   ❖

REALITY PROGRAMMING and Oprah heart-to-hearts
may not be the most sophisticated offering on the televised
menu, but neither are they the equivalent of junk food: a
guilty pleasure with no redeeming cognitive nourishment.
They engage the mind—and particularly the social mind—
far more rigorously than the worst shows of past decades.
People didn't gather at the water cooler to second-guess the
losing strategy on last night's *Battle of the Network Stars,*
but they'll spend weeks debating the tactical decisions and
personality tics of the *Apprentice* contestants. Consider this
one excerpt from an exchange on an unofficial *Apprentice* site:

> **KMJ179:** A person who is a loose cannon panics quite
> easily and makes hasty decisions without knowing the
> facts or realizes what is at stake. Loose cannons do not lis-
> ten to other people. Often times they will hear someone
> talking to them but they do not listen to what is being
> said. A loose cannon is someone who says one thing but
> turns around and does another thing on his or her own.
> I have dealt with loose cannons before and Troy is not a
> loose cannon by any means. Where Bernie got that from
> I do not know. It may have been Troy's accent that both-
> ered the poor Bernie.

Ken NJ: I'm not defending Bernie, but merely providing my reasons so that you can see where I'm coming from in classifying Troy as a loose cannon. He was expected by Donald, his team mates and his TV audience to put in an honest days work for a honest days pay. Well, he didn't performed honestly and started the "hook or by crook" method with some false representations to clients in misleading them to bid by some undue influence. Any responsible executive seeing Troy's business tactics on-the-job would say this worker is a loose cannon because he can't conform to corporate policies and marches to his own tune. Even Bill who has observed own co-worker said he had serious questions about the way Troy goes about closing his deals.

KMJ179: I was surprised when Troy crossed the Ethical boundry and resorted to lying about the actual number of people interested in renting the place. He did not have to do that. Ireonically when Troy was up front with the potential second client about having the first client also interested and sitting in another office, Troy lost out. The second client felt like he was beeing hussled. In a way I could not blame the second client though. We are talking about a high lease price for one day and you are telling me that I am competing with someone else for the highest price. I would tell Troy to go jump in the Hudson. Troy was very professional and let the client go after thanking him for the opportunity to meet.

❖   ❖   ❖

Ken NJ: You just illustrated one incident of Troy's un-
acceptable method of doing business. I've seen used-car
salesperson with more style and honesty than Troy. The
other instance, I've posted about Troy pulling the Kwame
autograph sales in Planet Hollywood curbside in misleading patrons. The Better Business Bureau and the State
Consumer Agencies would be starting investigations on
such pattern of business practices. I've seen aggressive
sales people like Troy bankrupt profitable businesses
overnight where the courts awarded treble damages in
multimillion judgements. Troy is a live trip wire, just wait-
ing to blow up the company. That's NOT an understate-
ment in today's corporate governance.

It would probably take you a lifetime to read all the tran-
scripts of comparable debates, both online and off, that follow in the wake of these shows. The spelling isn't perfect,
and the grammar occasionally leaves something to be desired. But the level of cognitive engagement, the eagerness
to evaluate the show through the lens of personal experience
and wisdom, the tight focus on the contestants' motives and
character flaws—all this is remarkable. It's impossible to
imagine even the highbrow shows of yesteryear—much less
The Dukes of Hazard—inspiring this quantity and quality

of analysis. (There are literally hundreds of pages of equivalent commentary at this one fan site alone.) The unique cocktail that the reality genre serves up—real people, evolving rule systems, and emotional intimacy—prods the mind into action. You don't zone out in front of shows like *The Apprentice*. You play along.

The content of the game you're playing, admittedly, suffers from a shallow premise and a highly artificial environment. (Plus the show forces you to contemplate Donald Trump's comb-over on a regular basis, occasionally windblown.) This is another way in which the reality shows borrow their techniques from the video games: the content is less interesting than the cognitive work the show elicits from your mind. It's the collateral learning that matters.

Part of that collateral learning comes from the sheer number of characters involved in a show like *The Apprentice* or *Survivor*. Just as *The Sopranos* challenges the mind to follow multiple threads, the reality shows demand that we track multiple *relationships,* since the action of these shows revolves around the shifting feuds and alliances between more than a dozen individuals. This, too, activates a component of our emotional IQ, sometimes called our social intelligence: our ability to monitor and recall many distinct vectors of interaction in the population around us, to remember that Peter hates Paul, but Paul likes Peter, and both of them get along with Mary. This faculty is part of our primate heritage; our closest relatives, the chimpanzees, live in societies characterized by intricate political calculation between dozens of individuals. (Some anthropologists believe that the explosion in frontal lobe size experienced by *Homo sapiens* over the past million years was spurred by the need to assess densely interconnected social networks.) Environmental conditions can strengthen or weaken the brain's capacity for this kind of social mapping, just as it can for real-world mapping. A famous study by University College London found that London cabdrivers had, on average, larger regions in the brain dedicated to spatial memory than the ordinary Londoner. And veteran drivers had larger areas than their younger colleagues. This is the magic of the brain's plasticity: by executing a certain cognitive function again and again, you recruit more neurons to participate in the task. Social intelligence works the same way: spend more hours studying the intricacies of a social network, and your brain will grow more adept at tracking all those intersecting relationships.

Where media is concerned, this type of analysis is not adequately illustrated by narrative threads or a simple list of characters. It is better visualized as a network: a series of points connected by lines of affiliation. When we watch most reality shows, we are implicitly building these social network maps in our heads, a map not so much of plotlines as of attitudes: Nick has a thing for Amy, but Amy may just be using Nick; Bill and Kwame have a competitive friendship, and both think Amy is using Nick; no one trusts Omarosa, except Kwame, but Troy *really* doesn't trust Omarosa. This may sound like high school, but like many forms of emotional intelligence, the ability to analyze and recall the full range of social relationships in a large group is just as reliable a predictor of professional success as your SAT scores or your college grades. Thanks to our biological and cultural heritage, we live in large bands of interacting humans, and people whose minds are skilled at visualizing all the relationships in those bands are likely to thrive, while those whose minds have difficulty keeping track are invariably handicapped. Reality shows force us to exercise that social muscle in ways that would have been unimaginable on past game shows, where the primary cognitive skill tested was the ability to correctly guess the price of a home appliance, or figure out the right time to buy a vowel.

The trend toward increased social network complexity is not the exclusive province of reality television; many popular television dramas today feature dense webs of relationships that require focus and scrutiny on the part of the viewer just to figure out what's happening on the screen. Traditionally, the most intricate social networks on television have come in the form of soap operas, with affairs and betrayals and tortured family dynamics. So let's take as a representative example an episode from season one of *Dallas*. The social network at the heart of *Dallas* is ultimately the Ewing family: two parents, three children, two spouses.

A few regular characters orbit at the periphery of this con-
stellation: the farmhand Ray, the Ewing nemesis Cliff. Each
episode introduces a handful of characters who play a one-
time role in that week's plotline and then disappear from the
network. In this episode, "Black Market Baby," the primary
structure of the narrative is a double plot: the competition
between the two brothers to have a baby and give the family patriarch a long-overdue grandchild. Imagined purely in
narrative terms—along the lines of our *Sopranos* and *Hill
Street*—this would be a relatively simple structure: two plotlines bouncing back and forth, overlapping at a handful of
key moments. But viewed as a social network, it is a more
nuanced affair:

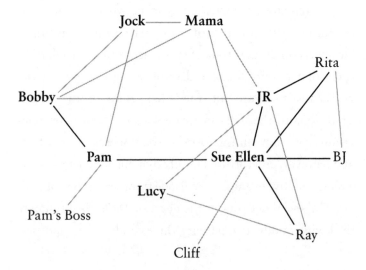

The lighter lines represent a social relationship that you must grasp to make sense of the episode's plot: you need to understand that the patriarch Jock doesn't approve of Pam's decision to go into the workforce and delay having a baby, just as you need to understand the longstanding rivalry between Bobby and JR in several crucial scenes with the entire family. The darker lines represent social relationships that trigger primary narrative events: when JR intervenes to pay the surrogate mother Rita to leave the state, thereby squelching Sue Ellen's adoption plan, or when Sue Ellen has a drunken night of passion with Ray.

Most of us don't think of these social networks in explicitly spatial terms while we watch TV, of course, but we do build working models of the social universe as we watch. The visualizations help convey in a glance how complex the universe is. And a glance is all you need to see—in the chart on page 112, of a season-one episode of the FOX series *24*—that something profound has happened to the social complexity of the TV drama in the past thirty years.

Season one of *24* is ultimately a narrative web strung between four distinct families: the hero Jack Bauer and his wife and daughter; the family of the threatened senator, David Palmer; the family of the Serbian terrorist Victor Drazen; and the informal family of coworkers at the Central Terrorism Unit, where Bauer works. (This last functions as a family not just because they live in close quarters together, but also because the office dynamics include two

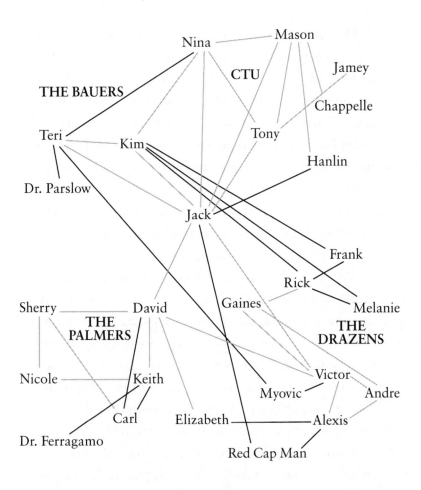

significant romantic dalliances.) Again, I have represented social connections that are relevant to the episode's plot in the lighter lines, and relationships crucial to the plot in darker lines. By every conceivable measure, *24* presents at least three times as complex a network as *Dallas*: the number of characters; the number of distinct groups; the connections between characters, and between groups; the number of relationships that are central to the episode's narrative. The social world of *Dallas* is that of an extended family: the primary players are direct relatives of one another, and the remaining characters have marginal roles. *24,* on the other hand, is closer to the scale of a small village, with four rival clans and dozens of links connecting them. Indeed, the social network of *24* mirrors the social network you frequently encounter in the small-town or estate novels of Jane Austen or George Eliot. The dialogue and description are more nuanced in those classic works, of course, but in terms of the social relationships you need to follow to make sense of the narrative, *24* holds its own.

Watch these two episodes of *Dallas* and *24* side by side and the difference is unavoidable. The social network of *Dallas* is perfectly readable within the frame of the episode itself, even if you haven't seen the show before and know nothing of its characters. The show's creators embed flashing arrows throughout the opening sequence—an extended birthday party for the family patriarch, Jock—that laboriously outline the primary relationships and tensions within the family. Keeping track of the events that follow requires almost no thought: the scenes are slow enough, and the narrative crutches obvious enough, that the modern television fan is likely to find the storylines sluggish and obvious. Watch *24* as an isolated episode and you'll be utterly baffled by the events, because they draw on such a complex web of relationships, almost all of which have been defined in previous installments of the series. Appropriately enough for a narrative presented in real time, 24 doesn't waste precious seconds explaining the back story; if you don't remember that Nina and Tony are having an affair, or that Jack and David collaborated on an assassination attempt against Drazen, then you'll have a hard time keeping up. The show doesn't cater to the uninitiated. But even if you *have* been following the season closely, you'll still find yourself straining to keep track of the plot, precisely because so many relationships are at play.

The map of 24's social network actually understates the cognitive work involved in parsing the show. As a conspiracy narrative—and one that features several prominent "moles"—each episode invariably suggests what we might call phantom relationships between characters, a social connection that is deliberately not shown onscreen, but that viewers inevitably ponder in their own minds. In this episode of 24, Jack Bauer's wife, Teri, suffers from temporary amnesia and spends some time under the care of a new character, Dr. Parslow, about which the viewer knows nothing. The show offers no direct connection to the archvillain, Victor Drazen, but in watching Parslow comfort Teri, you compulsively look for clues that might connect him to Drazen. (The same kind of scrutiny follows all the characters at

CTU, because of the mole plot.) In *24*, following the plot is not merely keeping track of all the dots that the show connects for you; the allure of the show also lies in weighing *potential* connections even if they haven't been deliberately mapped onscreen. Needless to say, *Dallas* marks all its social relationships with indelible ink; the shock of the "Who shot JR?" season finale lay precisely in the fact that a social connection—between JR and his would-be assassin—was for once *not* explicitly spelled out by the show.

Once again, the long-term trend of the Sleeper Curve is clear: one of the most complex social networks on popular television in the seventies looks practically infantile next to the social networks of today's hit dramas. The modern viewer who watches *Dallas* on DVD will be bored by the content—not just because the show is less salacious than today's soap operas (which it is by a small margin) but because the show contains far less *information* in each scene. With *Dallas,* you don't have to think to make sense of what's going on, and not having to think is boring. *24* takes the opposite approach, layering each scene with a thick network of affiliations. You have to focus to follow the plot, and in focusing you're exercising the part of your brain that maps social networks. The content of the show may be about revenge killings and terrorist attacks, but the collateral learning involves something altogether different, and more nourishing. It's about relationships.

THE INTERNET

VIEWERS WHO GET LOST in *24*'s social network have a resource available to them that *Dallas* viewers lacked: the numerous online sites and communities that share information about popular television shows. Just as *Apprentice* viewers mulled Troy's shady business ethics in excruciating detail, *24* fans exhaustively document and debate every passing glance and brief allusion in the series, building detailed episode guides and lists of Frequently Asked Questions. One Yahoo! site featured at the time of this writing more than forty thousand individual posts from ordinary viewers, contributing their own analysis of last night's episode, posting questions about plot twists, or speculating on the upcoming season. As the shows have complexified, the resources for making sense of that complexity have multiplied as well. If you're lost in *24*'s social network, you can always get your bearings online.

All of which brings us to another crucial piece in the puzzle of the Sleeper Curve: the Internet. Not just because the online world offers resources that help sustain more complex programming in other media, but because the process of acclimating to the new reality of networked communications has had a salutary effect on our minds. We do well to remind ourselves how quickly the industrialized world has embraced the many forms of participatory electronic media—from e-mail to hypertext to instant messages and blogging. Popular audiences embraced television and the cinema in comparable time frames, but neither required the learning curve of e-mail or the Web. It's one thing to adapt your lifestyle to include time for sitting around watching a moving image on a screen; it's quite another to learn a whole new language of communication and a small army of software tools along with it. It seems almost absurd to think of this now, but when the idea of hypertext documents first entered the popular domain in the early nineties, it was a distinctly avant-garde idea, promoted by an experimentalist literary fringe looking to explode the restrictions of the linear sentence and the page-bound book. Fast forward less than a decade, and something extraordinary occurs: exploring nonlinear document structures becomes as second nature as dialing a phone for hundreds of millions—if not billions—of people. The mass embrace of hypertext is like the *Seinfeld* "Betrayal" episode: a cultural form that was once exclusively limited to avant-garde sensibilities, now happily enjoyed by grandmothers and third-graders worldwide.

I won't dwell on this point, because the premise that increased interactivity is good for the brain is not a new one. (A number of insightful critics—Kevin Kelly, Douglas Rushkoff, Janet Murray, Howard Rheingold, Henry Jenkins—have made variations on this argument over the past decade or so.) But let me say this much: The rise of the Internet has challenged our minds in three fundamental and related ways: by virtue of being participatory, by forcing users to learn new interfaces, and by creating new channels for social interaction.

Almost all forms of online activity sustained are participatory in nature: writing e-mails, sending IMs, creating photo logs, posting two-page analyses of last night's *Apprentice* episode. Steve Jobs likes to describe the difference between television and the Web as the difference between lean-back and sit-forward media. The networked computer makes you lean in, focus, engage, while television encourages you to zone out. (Though not as much as it used to, of course.) This is the familiar interactivity-is-good-for-you argument, and it's proof that the conventional wisdom is, every now and then, actually wise.

There was a point several years ago, during the first wave of Internet cheerleading, when it was still possible to be a skeptic about how participatory the new medium would turn out to be. Everyone recognized that the practices of composing e-mail and clicking on hyperlinks were going to be mainstream activities, but how many people out there were ultimately going to be interested in publishing more extensive material online? And if that turned out to be a small number—if the Web turned out to be a medium where most of the content was created by professional writers and editors—was it ultimately all that different from the previous order of things?

The tremendous expansion of the blogging world over

the past two years has convincingly silenced this objection. According to a 2004 study by the Pew Charitable Trust, more than 8 million Americans report that they have a personal weblog or online diary. The wonderful blog-tracking service Technorati reports that roughly 275,000 blog entries are published in the average day—a tiny fraction of them authored by professional writers. After only two years of media hype, the number of active bloggers in the United States alone has reached the audience size of prime-time network television.

So why were the skeptics so wrong about the demand for self-publishing? Their primary mistake was to assume that the content produced in this new era would look like old-school journalism: op-ed pieces, film reviews, cultural commentary. There's plenty of armchair journalism out there, of course, but the great bulk of personal publishing is just that, *personal*: the online diary is the dominant discursive mode in the blogosphere. People are using these new tools not to opine about social security privatization; they're using the tools to talk about their lives. A decade ago Douglas Rushkoff coined the phrase "screenagers" to describe the first generation that grew up with the assumption that the images on a television screen were supposed to be manipulated; that they weren't just there for passive consumption. The next generation is carrying that logic to a new extreme: the screen is not just something you manipulate, but something you project your identity onto, a place to work through the story of your life as it unfolds.

To be sure, that projection can create some awkward or unhealthy situations, given the public intimacy of the online diary, and the potential for identity fraud. But every new technology can be exploited or misused to nefarious ends. For the vast majority of those 8 million bloggers, these new venues for self-expression have been a wonderful addition to their lives. There's no denying that the content of your average online diary can be juvenile. These diaries are, after all, frequently created by juveniles. But thirty years ago those juveniles weren't writing novels or composing sonnets in their spare time; they were watching *Laverne & Shirley*. Better to have minds actively composing the soap opera of their own lives than zoning out in front of someone else's.

The Net has actually had a positive lateral effect on the tube as well, in that it has liberated television from attempting tasks that the medium wasn't innately well suited to perform. As a vehicle for narrative and first-person intimacy, television can be a delightful medium, capable of conveying remarkably complex experiences. But as a source of information, it has its limitations. The rise of the Web has enabled television to offload some of its information-sharing responsibilities to a platform that was designed specifically for the purposes of sharing information. This passage from Postman's *Amusing Ourselves to Death* showcases exactly how much has changed over the past twenty years:

Television . . . encompasses all forms of discourse. No one goes to a movie to find out about government policy or the latest scientific advance. No one buys a record to find out the baseball scores or the weather or the latest murder. . . . But everyone goes to television for all these things and more, which is why television resonates so powerfully throughout the culture. Television is our culture's principal mode of knowing about itself.

No doubt in total hours television remains the dominant medium in American life, but there is also no doubt that the Net has been gaining on it with extraordinary speed. If the early adopters are any indication, that dominance won't last for long. And for the types of knowledge-based queries that Postman describes—looking up government policy or sports scores—the Net has become the first place that people consult. Google is *our* culture's principal way of knowing about itself.

The second way in which the rise of the Net has challenged the mind runs parallel to the evolving rule systems of video games: the accelerating pace of new platforms and software applications forces users to probe and master new environments. Your mind is engaged by the interactive content of networked media—posting a response to an article online, maintaining three separate IM conversations at the same time—but you're also exercising cognitive muscles interacting with the *form* of the media as well: learning the tricks of a new e-mail client, configuring the video chat software properly, getting your bearings after installing a new operating system. This type of problem-solving can be challenging in an unpleasant way, of course, but the same can be said for calculus. Just because you don't like troubleshooting your system when your browser crashes doesn't mean you aren't exercising your logic skills in finding a solution. This extra layer of cognitive involvement derives largely from the increased prominence of the interface in digital technology. When new tools arrive, you have to learn what they're good for, but you also have to learn the rules that govern their use. To be an accomplished telephone user, you needed to grasp the essential utility of being able to have real-time conversations with people physically removed from you, *and* you had to master the interface of the telephone device itself. That same principle holds true for digital technologies, only the interfaces have expanded dramatically in depth and complexity. There's only so much cognitive challenge at stake in learning the rules of a rotary dial phone. But you could lose a week exploring all the nooks and crannies of Microsoft Outlook.

Just as we saw in the world of games, learning the intricacies of a new interface can be a genuine pleasure. This is a story that is not often enough told in describing our evolving relationship with software. There is a kind of exploratory wonder in downloading a new application, and meandering through its commands and dialog boxes, learning its tricks by feel. I've often found certain applications are more fun to explore the first time than they actually are to use—because in the initial exploration, you can delight in features that are clever without being terribly helpful. This sounds like something only a hardened tech geek would say, but I suspect the feeling has become much more mainstream over the past few years. Think of the millions of ordinary music fans who downloaded Apple's iTunes software: I'm sure many of them enjoyed their first walk through the application, seeing all the tools that would revolutionize the way they listened to music. Many of them, I suspect, eschewed the manual altogether, choosing to probe the application the way gamers investigate their virtual worlds: from the inside. That probing is a powerful form of intellectual activity—you're learning the rules of a complex system without a guide, after all. And it's all the more powerful for being fun.

Then there is the matter of social connection. The other concern that Net skeptics voiced a decade ago revolved around a withdrawal from public space: yes, the Internet might connect us to a new world of information, but it would come at a terrible social cost, by confining us in front of barren computer monitors, away from the vitality of genuine communities. In fact, nearly all of the most hyped developments on the Web in the past few years have been tools for augmenting social connection: online personals, social and business network sites such as Friendster, the Meetup.com service so central to the political organization of the 2004 campaign, the many tools designed to enhance conversation between bloggers—not to mention all the handheld devices that we now use to coordinate new kinds of real-world encounters. Some of these tools create new modes of communication that are entirely digital in nature (the cross-linked conversations of bloggers). Others use the networked computer to facilitate a face-to-face encounter (as in Meetup). Others involve a hybrid dance of real and virtual encounters, as in the personals world, where flesh-and-blood dates usually follow weeks of online flirting. Tools like Google have fulfilled the original dream of digital machines becoming extensions of our memory, but the new social networking applications have done something that the visionaries never imagined: they are augmenting our people skills as well, widening our social networks, and creating new possibilities for strangers to share ideas and experiences.

Television and automobile society locked people up in their living rooms, away from the clash and vitality of public space, but the Net has reversed that long-term trend. After a half-century of technological isolation, we're finally learning new ways to connect.

FILM

HAVE THE MOVIES UNDERGONE an equivalent transformation? The answer to that is, I believe, a qualified yes. The obvious way in which popular film has grown more complex is visual and technological: the mesmerizing special effects; the quicksilver editing. That's an interesting development, and an entertaining one, but not one that is likely to have a beneficial effect on our minds. Do we see the same growing narrative complexity, the same audience "filling in" that we see in television shows today? At the very top of the box office list, there is some evidence of the Sleeper Curve at work. For a nice apples-to-apples comparison, contrast the epic scale and intricate plotting of the *Lord of the Rings* trilogy to the original *Star Wars* trilogy. Lucas borrowed some of the structure for *Star Wars* from Tolkien's novels, but in translating them into a blockbuster space epic, he simplified the narrative cosmology dramatically. Both share a clash between darkness and light, of course, and the general structure of the quest epic. But the particulars are radically different. By each crucial measure of complexity—how many narrative threads you're forced to follow, how much background information you need to interpret on the fly—*Lord of the Rings* is several times more challenging than *Star Wars*. The easiest way to grasp this is simply to review the number of characters who have active threads associated with them, characters who affect the plot in some important way, and who possess a biographical story that the film conveys. *Star Wars* contains roughly ten:

Luke Skywalker
Han Solo
Princess Leia Organa
Grand Moff Tarkin
Ben Obi-Wan Kenobi
C-3PO
R2-D2
Chewbacca
Darth Vader

Lord of the Rings, on the other hand, forces you to track almost three times as many:

Everard Proudfoot
Sam Gamgee
Sauron
Boromir
Galadriel
Legolas Greenleaf
Pippin
Celeborn
Gil-galad

Bilbo Baggins
Gandalf
Saruman
Lurtz
Elendil
Aragorn
Haldir
Gimli
Gollum
Arwen
Elrond
Frodo Baggins

The cinematic Sleeper Curve is most pronounced in the genre of children's films. The megahits of the past ten years—*Toy Story; Shrek; Monsters, Inc.;* and the all-time money-making champ, *Finding Nemo*—follow far more intricate narrative paths than earlier films like *The Lion King, Mary Poppins,* or *Bambi.* Much has been written about the dexterity with which the creators of these recent films build distinct layers of information into their plots, dialogue, and visual effects, creating a kind of hybrid form that dazzles children without boring the grownups. (*Toy Story,* for instance, harbors an armada of visual references to other movies—*Raiders of the Lost Ark, The Right Stuff, Jurassic Park*—that wouldn't be out of place in a *Simpsons* episode.) But the most significant change in these recent films is structural.

Take as a representative comparison the plots of *Bambi* (1942), *Mary Poppins* (1964), and *Finding Nemo* (2002). Set aside the question of the life lessons imparted by these films—they are all laudable, of course—and focus instead on the number of distinct characters in each film who play an integral role in the plot, characters who are presented with some biographical information, who develop or change over the course of the film. (Characters with a "story arc," as screenwriting jargon has it.) All three films contain a family unit at their core: Bambi and Flower, the Bankses, Nemo and his widowed father. They also feature one or two main sidekicks who complement the family unit: Thumper, Mary Poppins and Bert, the amnesiac Dory. But beyond those shared characteristics, the plots diverge dramatically. *Bambi*'s plot revolves almost exclusively around those central three individuals; *Mary Poppins* introduces about five additional characters who possess distinct story arcs and biographical information (Bert the chimney sweep, the laughing uncle, the bank president). To follow *Nemo*'s plot, however, you have to keep track of almost twenty unique personalities: Nemo's three school chums and their teacher; the three recovering sharks including Bruce, who "never had a father"; the six fish in the aquarium, led by Gill, whose scarred right side bonds him to Nemo with his weak left fin; Crush, the surfer-dude turtle; Nigel the pelican; the aquarium-owning dentist and his evil niece. Add to that a parade of about ten oceanographic cameos: whales, lobsters, jellyfish—all of which play instrumental roles in the narratives without having clearly defined personalities. As the father of a three-year-old, I can testify personally that you can watch Nemo dozens of times and still detect new information with each viewing, precisely because the narrative floats so many distinct story arcs at the same time. And where the child's mind is concerned, each viewing is training him or her to hold those multiple threads in consciousness, a kind of mental calisthenics.

To see the other real explosion in cinematic complexity, you have to look to the mid-list successes, where you will find significant growth in films built around fiendishly complex plots, demanding intense audience focus and analysis just to figure out what's happening on the screen. I think of this as a new microgenre of sorts: the mind-bender, a film designed specifically to disorient you, to mess with your head. The list includes *Being John Malkovich, Pulp Fiction, L.A. Confidential, The Usual Suspects, Memento, Eternal Sunshine of the Spotless Mind, Run Lola Run, Twelve Monkeys, Adaptation, Magnolia,* and *Big Fish.* (You might add *The Matrix* to this list, since its genius lay in cleverly implanting the mind-bender structure within a big-budget action picture.)

Some of these films challenge the mind by creating a thick network of intersecting plotlines; some challenge by withholding crucial information from the audience; some by inventing new temporal schemes that invert traditional relationships of cause and effect; some by deliberately blurring the line between fact and fiction. (All of these are classic techniques of the old cinematic avant-garde, by the way.) There are antecedents in the film canon, of course: some of the seventies conspiracy films, some of Hitchcock's psychological thrillers. But the mind-benders have truly flowered as a genre in the past ten years—and done remarkably well at the box office too. Most of the films cited above made more than $50 million from box-office receipts alone, and all of them made money for their creators—despite their reliance on narrative devices that might have had them consigned to the art house thirty years ago.

But elsewhere in the world of film, the trends are less dramatic. At the top of the box office charts, I think it's fair to say that *Independence Day* is no more complex than *E.T.*; nor is *The Sixth Sense* more challenging than *The Exorcist*. Hollywood still churns out a steady diet of junk films targeted at teens that are just as simple and formulaic as they were twenty years ago. Why, then, does the Sleeper Curve level off in the world of film?

I suspect the answer is twofold. First, narrative film is an older genre than television or games. The great explosion of cinematic complexity happened in the first half of the twentieth century, in the steady march from the trompe l'oeil and vaudeville diversions of the first movies through *Birth of a Nation* and *The Jazz Singer* all the way to *Citizen Kane* and *Ben-Hur*. As narrative cinema evolved as a genre, and as audiences grew comfortable with that evolution, the form grew increasingly adventurous in the cognitive demands it made on its audience—just as television and games have done over the past thirty years. But film has historically confronted a ceiling that has reined in its complexity, because its narratives are limited to two to three hours. The television dramas we examined tell stories that unfold over multiple seasons, each with more than a dozen episodes. The temporal scale for a successful television drama can be more than a hundred hours, which gives the storylines time to complexify, and gives the audience time to become familiar with the many characters and their multiple interactions. Similarly, the average video game takes about forty hours to play, the complexity of the puzzles and objectives growing steadily over time as the game progresses. By this standard, your average two-hour Hollywood film is the equivalent of a television pilot or the opening training sequence of a video game: there are only so many threads and subtleties you can introduce in that time frame. It's no accident that the most complex blockbuster of our era—the *Lord of the Rings* trilogy—lasts more than ten hours in its uncut DVD version. In the recipe for the Sleeper Curve, the most crucial ingredient is also the simplest one: time.

*   *   *

THE SLEEPER CURVE charts a trend in the culture: popular entertainment and media growing more complex over time. But I want to be clear about one thing: The Sleeper Curve does not mean that *Survivor* will someday be viewed as our *Heart of Darkness,* or *Finding Nemo* our *Moby-Dick.* The conventional wisdom the Sleeper Curve undermines is *not* the premise that mass culture pales in comparison with High Art in its aesthetic and intellectual riches. Some of the long-form television dramas of recent years may well find their way into some kind of canon years from now, along with a few of the mind-benders. Games will no doubt develop their own canon, if they haven't already. But that is another debate. The conventional wisdom that the Sleeper Curve *does* undermine is the belief that things are getting worse: the pop culture is on a race to the bottom, where the cheapest thrill wins out every time. That's why it's important to point out that even the worst of today's television—a show like *The Apprentice,* say— doesn't look so bad when measured against the dregs of television past. If you assume there will always be a market for pulp, at least the pulp on *The Apprentice* has some connection to people's real lives: their interoffice rivalries, their battles with the shifting ethics and sexual politics of the corporate world. It's not the most profound subject matter in the history of entertainment, but compared with the pabulum of past megahits—compared with *Mork & Mindy* or *Who's the Boss?*—it's pure gold.

But in making this comparative argument, some might say I have set the bar too low. Perhaps the general public's appetite for pulp entertainment is not a sociological constant. If you think that the ecosystem of television will always serve up shows that exist on a spectrum of quality—some trash and some classics, and quite a bit in the middle—then it's a good sign when the trash seems to be getting more mentally challenging as the medium evolves. But if it's possible to avoid the trash altogether—a nation of PBS viewers—then we shouldn't be thankful for programs whose saving grace is solely that they aren't quite as dumb as the shows used to be.

When people hold out the possibility of such a cultural utopia, they often point to the literary best-seller lists of yesteryear, which allegedly show the masses devouring works of great intricacy and artistic merit. The classic case of highbrow erudition matched with popular success is Charles Dickens, who for a stretch of time in the middle of the nineteenth century was the most popular author writing in the English language, and also (with the possible exception of George Eliot) the most innovative. If the Victorians were willing to line up en masse to read *Bleak House*—with its thousand pages and byzantine plot twists, not to mention its artistic genius—why should we settle for *The Apprentice*?

It is true that Dickens's brilliance lay at least partially in his ability to expand the formal range of the novel while simultaneously building a mass audience eager to follow along. Indeed, Dickens helped to invent some of the essential conventions of mass entertainment—large groups of strangers united by a shared interest in a serialized narrative—that we now take for granted. That he managed to create enduring works of art along the way is one of the miracles of literary history, though of course it took the Cultural Authorities nearly a century to make him an uncontested member of the literary canon, partially because his novels had been tainted by their commercial success, and partially because Dickens's comic style made his novels appear less serious than those of his contemporaries.

So if Dickens could juggle Great Art and Mass Audience, why should we tolerate some of the lesser creatures that populate the high end of the Nielsen ratings today? The answer, I believe, is that the definition of a "mass success" has changed since Dickens's time. On average, Dickens sold around 50,000 copies of the serialized versions of his novels, during a time in which the British population was roughly 20 million. Had Dickens's potential audience been the size of the United States today—280 million people—he would have sold something like 800,000 copies of his first-run novels. The most innovative shows on television today—*The West Wing, 24, The Simpsons, The Sopranos*—often attract between 10 and 15 *million* viewers. So by this measure, *West Wing* is roughly twenty times more "mass" than Dickens was, even though Dickens had no mass media rivals for his audience's attention—no television or radio or cinema to compete with. It's no wonder

Dickens was able to persuade his readers to keep up with his rhetorical innovations. In his day, Dickens had the per capita audience that would today tune in for a Masterpiece Theatre airing of *Bleak House*. His audience was mass by Victorian standards; no genuinely literary author had attracted that many readers before. But by modern standards, he was writing for the elite.

Dickens may not have been a mass author by modern standards, but you needn't look far to find an example of truly mass cultural successes that are simultaneously the most complex and nuanced in their field. Violent video games like *Quake* or *Doom* tend to dominate the mainstream media discussion of gaming, but the fact is the shooter games are rarities on the gaming best-seller lists. The two genres that historically have dominated the charts are both forms of complex simulation: either sport sims, or GOD games like *SimCity* or *Age of Empires*. The most popular game of all time is the domestic saga *The Sims*. (The closest thing you'll see to a violent exchange in *The Sims* is when one of your virtual characters can't pay the monthly bills.) The sports simulations have reached a level of intricacy that makes the dice-baseball games I explored as a child look like tic-tac-toe—not just in their near-photorealistic graphics, but in the player's ability to control and model the most microscopic aspect of the game. Sega's *2K3* baseball simulator gives you an entire organization to general manage: trading players, nurturing minor leaguers, negotiating salaries and free agents. (This is not, incidentally, a universe of pure numbers. Emotions factor as well. Bench a highly paid prima donna for a few days, and his productivity will diminish, just as it will on the real-world diamond.) As for the social and historical simulations, just think back to my nephew learning about the effects of industrial taxes while playing *SimCity*. The violent games may generate the most outrage, but the games that people reliably line up to buy are the ones that require the most thinking. Somehow in this age of attention deficit disorder and instant gratification, in this age of gratuitous violence and cheap titillation, the most intellectually challenging titles are also the most popular. And they're growing more challenging with each passing year.

*   *   *

So THIS is the landscape of the Sleeper Curve. Games that force us to probe and telescope. Television shows that require the mind to fill in the blanks, or exercise its emotional intelligence. Software that makes us sit forward, not lean back. But if the long-term trend in pop culture is toward increased complexity, is there any evidence that our brains are reflecting that change? If mass media is supplying an increasingly rigorous mental workout, is there any empirical data that shows our cognitive muscles growing in response?

In a word: yes.

PART TWO

*   *   *

And Nietzsche, with his theory of eternal recurrence. He said that the life we lived we're going to live over again the exact same way for eternity. Great. That means I'll have to sit through the Ice Capades again.

—WOODY ALLEN

IN THE LATE SEVENTIES, an American philosopher and longtime civil-rights activist named James Flynn began investigating the history of IQ scores, in an attempt to refute studies published by controversial scholar Arthur Jensen, whose work later influenced the even more controversial book *The Bell Curve*. Jensen's research had uncovered an alleged gap between white and black IQ scores, a gap that wasn't attributable to differences in education or economic upbringing. Despite his lack of professional training in the field, Flynn decided to throw himself into the fray and prove that IQ tests were more culturally biased than Jensen had believed, thus making the racial IQ gap a byproduct of history not biology. Flynn's investigation led him to military records that clearly showed a dramatic increase in African-American IQ scores over the past half century, a trend that initially seemed to support his argument against Jensen: As African-Americans were granted greater access to the educational system, their IQ scores improved accordingly.

But as Flynn sifted through the data, he found something that challenged his expectations. Black scores were rising, to be sure. But white scores were rising almost as fast. Across the board, irrespective of class or race or education, Americans were getting smarter. Flynn was able to quantify the shift: in forty-six years, the American people had gained 13.8 IQ points on average.

The trend had gone unnoticed for so long because the IQ establishment routinely normalized the exams to ensure that a person of average intelligence scored 100 on the test. So every few years, they'd review the numbers and tweak the test to ensure that the median score was 100. Without realizing it, they were slowly but reliably increasing the difficulty of the test, as though they were ramping up the speed of a treadmill. If you looked exclusively at the history of the scores themselves, IQ seemed to be running in place, unchanged over the past century. But if you factored in the mounting challenge presented by the tests themselves, the picture changed dramatically: the test-takers were getting smarter.

Many of you may hold the opinion that IQ has been debunked by recent developments in brain science and sociology, and to a certain extent it has. That debunking has taken two primary forms: IQ has been shown to be more vulnerable to environmental conditions than its original "innate intelligence" billing indicated; and the intelligence that the IQ tests measure has been shown to reflect only part of the spectrum of human intelligence. But those objections—true as they may be—do not undermine the trend described by the Flynn Effect in any way. In fact, they may make it more interesting.

Clearly there are multiple forms of intelligence, only some of which are measured by IQ tests: emotional intelligence, for one, is entirely ignored by all traditional IQ metrics. And the Flynn Effect offers what many consider incontrovertible evidence that IQ is profoundly shaped by environment, since genetics alone can't explain such a dramatic rise in such a short amount of time. So when critics object to the practice of comparing individual or group IQs—as in *The Bell Curve*'s observation that African-Americans have, on average, lower IQs than those of white Americans—their objections have real merit: because IQ isn't the only gauge of real-world intelligence, and because differences in IQ may be due largely to environmental factors. Thus, IQ scores are less relevant in comparing the intelligence of, say, different ethnic groups—or even different candidates for college admission.

So why are IQ scores relevant to the Sleeper Curve? Because differences *between* generations don't pose the same problems that differences *within* generations do. When you look at a snapshot of black and white IQ tests from 1975, explaining the difference between those scores is a necessarily murky affair: each group possesses different combinations of genes *and* different environments. But when you look at IQ scores across generations, the picture gets clearer. Whatever genetic differences may exist between groups disappear, because you're looking at the average IQ of the entire society. The gene pool hasn't changed in a generation, and yet the scores have gone up. Some environmental factor (or combination of factors) must be responsible for the increase in the specific forms of intelligence that IQ measures: problem solving, abstract reasoning, pattern recognition, spatial logic.

Psychologists and social scientists and other experts in psychometrics have now had twenty years to study the Flynn Effect; while much debate remains about the ultimate causes behind the IQ increase, the existence of the trend itself is uncontested. IQs have been rising in most developed countries at an extraordinary clip over the past century: an average of 3 points per decade. A number of studies have suggested that the rate of increase is itself accelerating: average scores in the Netherlands, for instance, increased 8 points between 1972 and 1982. A few points may not sound like much, but the numbers quickly add up. Imagine this scenario: a person who tests in the top 10 percent of the United States in 1920 time-travels eighty years into the future and takes the test again. Thanks to the Flynn Effect, he would be in the bottom third for IQ scores today. Yesterday's brainiac is today's simpleton.

A small part of the Flynn Effect may be attributable to increased familiarity with intelligence tests themselves. But as Flynn points out, even if you take the exact same IQ test multiple times in a row, the benefits from that repeat exposure cap out at around 5 or 6 points. And the heyday of IQ testing was the middle of the twentieth century. Over the past thirty years, the rise in IQ scores has been accelerating, even as the administration of IQ tests has become less common.

Nor is the Flynn Effect likely to be the product of better nutrition. Adult height is famously sensitive to early diet, and indeed average heights have been on the rise for most of the past two centuries in the industrialized world. But in the United States and Europe the trend toward increased height leveled in the decades after World War II, presumably corresponding to a leveling off in the trend toward improved childhood nutrition. And yet the postwar period shows the most dramatic spike in IQ. If better nutrition were sharpening our brains, we would expect to see height increases running parallel to IQ increases. We would also expect to see improvements across the board in mental function, and not just the logic tests of IQ. But on tests that measure skills specifically taught in the classroom—math or history—U.S. students have been flatlining or worse for much of the past forty years. This suggests that improved education cannot be responsible for the Flynn Effect. For decades now, the recurring story about the U.S. educational system has long been its lagging test scores, numbers that are cited again and again whenever critics rail against failing public schools. They're right to complain, because those indices do measure skills that are important in real-world success, for both the individual and the society. But beneath those sorry numbers, a strangely encouraging trend continues: Where pure problem-solving is concerned, we're getting smarter.

If we're not getting these cognitive upgrades from our diets or our classrooms, where are they coming from? The answer should be self-evident by now. It's not the change in our nutritional diet that's making us smarter, it's the change in our *mental* diet. Think of the cognitive labor—and play—that your average ten-year-old would have experienced outside of school a hundred years ago: reading books when they were available, playing with simple toys, improvising neighborhood games like stickball and kick the can, and most of all doing household chores—or even working as a child-laborer. Compare that to the cultural and technological mastery of a ten-year-old today: following dozens of professional sports teams; shifting effortlessly from phone to IM to e-mail in communicating with friends; probing and telescoping through immense virtual worlds; adopting and troubleshooting new media technologies without flinching. Thanks to improved standards of living, these kids also have more time for these diversions than their ancestors did three generations before. Their classrooms may be overcrowded and their teachers underpaid, but in the world outside of school, their brains are being challenged at every turn by new forms of media and technology that cultivate sophisticated problem-solving skills.

Practically every family with young children has a running gag about how little Junior knows how to program the VCR while Mom and Dad with their advanced degrees can barely set the alarm clock. But I suspect we're too quick to write these skills off as mere superficial technical knowledge. The ability to take in a complex system and learn its rules on the fly is a talent with great real-world applicability; just like learning to read a chessboard, the content of the skill isn't as important as the general principles that underlie it. When your ten-year-old figures out how to consolidate all seven remote controls into a single unit, she's exercising problem-solving muscles with an insistence that rivals anything she's learning at school. You want your children fixing your home theater setup, not because they'll be able to use that skill working for Circuit City one day, but rather because there's a commendable structure to this kind of thinking.

The social psychologist Carmi Schooler sees the Flynn Effect as a reflection of environmental complexity:

The complexity of an individual's environment is defined by its stimulus and demand characteristics. The more diverse the stimuli, the greater the number of decisions required, the greater the number of considerations to be taken into account in making these decisions, and the more ill-defined and apparently contradictory the contingencies, the more complex the environment. To the degree that such an environment rewards cognitive effort, individuals should be motivated to develop their intellectual capacities and to generalize the resulting cognitive processes to other situations.

Environmental complexity is not limited to media, of course, but the characteristics that Schooler outlines describe precisely the contours of the Sleeper Curve: first, the emergence of media—like games and other interactive forms—that force decision-making at every turn; the increase in social and narrative complexity evident in television and some film; the intoxicating rewards of popular entertainment. All these forces working together create an environment likely to enhance problem-solving skills. Other forms of modern complexity may also be a factor here, of course: urban environments are, by Schooler's definition, more complex than rural ones, and so the industrial-age migration to the cities may play a role in the Flynn Effect. But most of the industrialized world underwent that migration before World War II; the post-war trend has been surburban flight. And so the most dramatic spike in IQ scores—the one witnessed over the past thirty years—is most likely being driven by something else.

*   *   *

THE LINK between the Flynn Effect and popular media is a hypothesis, but there are a number of reasons to think that more than a casual connection exists. As research into the Flynn Effect has deepened, three important tendencies have come to light, all of which parallel the developments in popular culture I've described over the preceding pages. The first is the general pattern itself: higher IQs mirroring the increased complexity of the culture. But in exploring the specifics of those IQ scores, researchers discovered a second trend in the data: the historical increase grew more dramatic the further the tests ventured from skills—like mathematic or verbal aptitude—that reflect educational background. The Flynn Effect is most pronounced on tests that assess what psychometricians call *g*, the index that offers the best approximation of "fluid" intelligence. Tests that measure *g* often do away with words and numbers, replacing them with questions that rely exclusively on images, testing the subject's ability to see patterns and complete sequences with elemental shapes and objects, as in this ex-

ample from the Raven Progressive Matrices test, which asks
you to fill the blank space with the correct shape from the
eight options below:

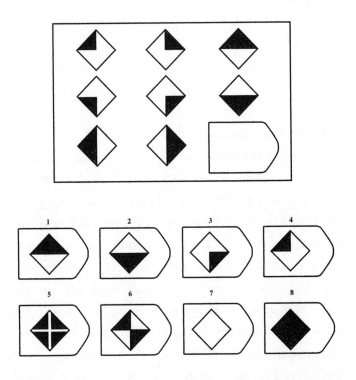

The centrality of the *g* scores to the Flynn Effect is telling.
If you look at intelligence tests that track skills influenced
by the classroom—the Wechsler vocabulary or arithmetic
tests, for instance—the intelligence boom fades from view;
SAT scores have fluctuated erratically over the past decades.
But if you look solely at unschooled problem-solving and

pattern-recognition skills, the progressive trend jumps into focus. There's something mysterious in these simultaneous trends: if *g* exists in a cultural vacuum, how can scores be rising at such a clip? And more puzzling, how can those scores be rising faster than other intelligence measures that do reflect education? The mystery disappears if you assume that these general problem-solving skills *are* influenced by culture, just not the part of culture that we conventionally associate with making people smarter. Their problem-solving skills are the result of the conditioning they get from interacting with popular culture that has grown more challenging over time. When you spend your leisure time interacting with media and technology that forces you to "fill in" and "lean forward," you're developing skills that will ultimately translate into higher *g* scores. (For those of you curious about your own skills, the correct answer to the Raven test question above is 8.)

Consider the kind of thinking you have to do to perform well on the Raven test. First, the information is presented in a visual language, not a textual one. You need—literally—to "fill in" the missing space and complete the sequence. You can't fill in by memorizing facts or having a large vocabulary; you have to do it by paying close attention to the grid, by detecting patterns in each object, by separating the relevant information from the irrelevant. You're presented, in effect, with a grid of potential clues that suggest what the missing box should contain; those clues are defined as a series of relationships: each shape connecting to other shapes in the grid in subtle ways. To solve this particular puzzle, you have to grasp that the essential relationships between the shapes run on both the vertical and horizontal axes, moving left to right and top to bottom, and involve adding the dark areas in the first two shapes together to create the proper coloration in the third shape. But the diagonal axes, for instance, are irrelevant. In this sense, there's an open-ended nature to the question: part of figuring out the solution lies in figuring out which elements of the question are pertinent and which are red herrings. If you ask someone to name the state capital of Missouri, or the square root of 128, there's no need to parse the question and determine which components are relevant or not: you either know the information by rote, or in the latter instance, you know the procedure for extracting a square root from a given number. The Raven grids, on the other hand, force you to separate the essential and the peripheral in the question itself.

This is exactly the kind of thinking that has become widespread in the popular media over the past few decades. Games, of course, rely heavily on this pattern recognition and deciphering; some puzzle games like *Tetris* even look like the Raven test. When you're mapping the complex relationships of *24* to figure out who the mole is, you're doing a social network rendition of the Raven grid: looking for patterns of behavior that reveal a hidden identity. When you're trying to figure out why your new e-mail client keeps crashing your PC, you're analyzing an array of potential clues—separating the essential from the peripheral—to figure out the underlying conflict. In all these activities, you have to analyze a complex tableau, build a working model of it in your head, and then make a decision. In the most basic sense, these different forms of media reward you for *solving* something.

The emphasis on abstract problem-solving in tests like the Raven originally stemmed from a desire to create tests that were free of cultural bias. It was better to ask people to mentally rotate rectangles in their heads than it was to ask them to analyze paragraphs about the Founding Fathers, because there were invariably culturally endowed facts and skills in the latter that favored certain demographic groups over others. For a while, this approach probably worked, precisely because there were no cultural groups that placed a disproportionate emphasis on mentally rotating a rectangle 270 degrees. But a few years ago, all of that started to change. A new group appeared that compulsively rotated rectangles all day long, that literally rotated rectangles in their sleep. But this group didn't break down into the usual economic or racial divisions. These weren't prep school elites, or Japanese-Americans, or the urban underclass. They were kids who played *Tetris.*

One other tendency in the history of IQ mirrors the trends in popular culture we've explored. The Flynn Effect is most pronounced in the low-to-mid range of intelligence scores.

At the very high end of IQ—the top 2 or 3 percentile—the curve levels off. Moderately intelligent people today are much smarter—at least where *g* is concerned—than moderately intelligent people were a hundred years ago. But a Mensa member today with a 150 IQ wouldn't be able to run circles around a genius from 1900. This is precisely the result we would expect to see if lowbrow culture and middlebrow culture are a driving force behind the Flynn Effect: while a person of moderate intelligence will have his or her pattern recognition talents sharpened by playing *Zelda* or studying the plotlines of *24*, a genius would probably require more challenging fare to improve his or her skills. Spending a week reviewing multiplication flash cards will decidedly improve the math skills of a fourth-grader, but it probably won't improve the skills of a college physics major. The same goes for popular media and *g*. The Sleeper Curve shows that the popular culture is growing more complex, yet it is not sufficiently complex to challenge the most gifted minds, which is why the geniuses aren't getting any smarter. What has changed is the cognitive workout that mass culture offers the rest of us.

Science is only beginning to understand what that workout actually entails. While many studies have analyzed the impact of television violence on behavior—with no clear consensus either way—the *positive* mental impact of contemporary media has not been widely examined. But a handful of recent studies have looked at the effect of playing video games on visual intelligence and memory. One study at the University of Rochester asked subjects to perform a series of quick visual recognition tests, picking out the color of a letter or counting the number of objects on a screen. The test was not as intricate as the Raven matrices, but it was more time-sensitive. Regular gamers consistently outperformed non-gamers on all the skills measured by the study. The researchers also debunked the premise that visually intelligent people are more likely to be attracted to video games in the first place. They had a group of non-players spend a week immersed in *Tetris* and the World War II game *Medal of Honor,* and found that this group's skills on the visual test improved as well. Games were literally making them perceive the world more clearly.

Another recent study looked at three distinct groups of white-collar professionals: hard-core gamers, occasional gamers, and non-gamers. The results contradict nearly all the received ideas about the impact of games: the gaming population turned out to be consistently more social, more confident, and more comfortable solving problems creatively. They also showed no evidence of reduced attention spans compared with non-gamers.

These early studies are tantalizing, but they are only the beginning. Because we have lived so long under the dumbing-down hypothesis, because we have been inclined to evaluate these new cultural forms as debased versions of older forms, we have very little data on positive cognitive impact, beyond the macro trend of the Flynn Effect. My hope is that we are beginning to appreciate some of these virtues, and that we will soon see research into the impact of gaming on probing and telescoping in complex environments, or the relationship between following television dramas and our ability to map social networks. Until that time, the most compelling evidence for the Sleeper Curve is financial: games and narratives that were too intricate for mass audiences thirty years ago now regularly attract millions of willing enthusiasts. Clearly *something* has changed in the minds of all those people that keeps them from being unpleasantly disoriented by these experiences. It's time we tried to figure out exactly what that something is.

Flynn's own position on the trend he discovered is itself iconoclastic. On the one hand, he remains convinced of the original insight that drove him into this line of inquiry nearly three decades ago: IQ is far more vulnerable to environmental conditions than previously believed. (In 2001, he coauthored a fascinating paper on the interaction of culture and genetics that explained why previous studies showing high rates of heritability for IQ neglected environmental factors.) And if environmental factors are responsible for the increase in IQ over the past fifty years, the next logical question is: What has changed in the environment over that time? In the industrialized world, where the Flynn Effect has been most pronounced, the answer is simple: Media and technology. Our diets haven't improved; our schools are more crowded and less endowed; our living environments are increasingly suburban. But the media and technology that our minds grapple with every day have grown at an exponential rate over that period, in both the complexity of the individual object and the diversity of the overall ecosystem. The mind is more challenged following the plot of *24* than the plot of *Dragnet*, and the mind is more challenged mastering the dozens of new media forms—games, hypertext, instant messaging, TiVo—that constitute mainstream culture today.

Yet Flynn has a twist. He sees the Flynn Effect undermining not only the genetics of IQ, but also the correlation between IQ and real-world intelligence. "Just as an elite with a massive IQ advantage should radically outperform the rest of its generation," he writes, "so a generation with a massive IQ gain should radically outperform its predecessors. . . . The result should be a cultural renaissance too great to be overlooked." And yet we see no evidence of "a dramatic increase in genius or mathematical and scientific discovery during the present generation." If IQs are improving but the culture isn't, then IQ must not be as useful a measure of intelligence as its supporters believe.

This is a book about a popular culture and not the history of science, so I'll leave Flynn's claims about the state of mathematical and scientific discovery for others to dispute in more detail. (Suffice it to say that the age of brain imaging, genome mapping, and the microchip stacks up nicely against past eras—particularly when you look at the sheer number of individuals contributing groundbreaking work, as opposed to the isolated geniuses of the past.) But in focusing on the idea of cultural renaissance, Flynn is looking at the outer edge of the bell curve, among the savants and visionaries. As we've seen, the Flynn Effect is most pronounced in the middle regions: the average person has seen the most dramatic IQ increase over the past decades. And the average person, like it or not, doesn't trigger scientific revolutions or cultural renaissances. The sharpening of his mind can't be measured at the extremes of intellectual achievement. Instead, we should detect that improvement somewhere else, in the everyday realm of managing more complex forms of technology, mastering increasingly nuanced narrative structures—even playing more complicated video games. We should detect that improvement in the realm of the Sleeper Curve. Flynn was right to say we should expect to find a cultural renaissance if the general rise in IQ truly measured an increase in intelligence. It's just that the culture turned out to be mass, not elite.

*　　*　　*

IF RISING IQs and the TV ratings suggest that the Sleeper Curve is having a beneficial impact on our mental faculties, one crucial question remains. Why is this tendency toward increased complexity happening in the first place? It is a truth nearly universally acknowledged that pop culture caters to our base instincts; mass society dumbs down and simplifies; it races to the bottom. The rare flowerings of "quality programming" only serve to remind us of the overall downward slide. But no matter how many times this refrain is belted out, it doesn't get any more accurate. As we've seen, precisely the opposite seems to be happening: the secular trend is toward greater cognitive demands, more depth, more participation. And if you accept that premise, you're forced then to answer the question: *Why?* For decades, the race to the bottom served as a kind of Third Law of Thermodynamics for mass society: all other things being equal, pop culture will decline into simpler forms. But if entropy turns out not to govern the world of mass society—if our entertainment is getting smarter after all—we need a new model to explain the trend.

That model is a complex, layered one. The forces driving the Sleeper Curve straddle three different realms of experience: the economic, the technological, and the neurological. Part of the Sleeper Curve reflects changes in the market forces that shape popular entertainment; part emanates from long-term technological trends; and part stems from deep-seated appetites in the human brain.

The Sleeper Curve is partly powered by the force of repetition. Over the past twenty years, a fundamental shift has transformed the economics of popular entertainment: original runs are now less lucrative than repeats. In the old days of television and Hollywood, the payday came from your initial airing on network or your first run at the box office. The aftermarkets for content were marginal at best. But the mass adoption of the VCR, and cable television's hunger for syndicated programming, has turned that equation on its head. In 2003, for the first time, Hollywood made more money from DVD sales than it did from box office receipts. Television shows repurposed as DVDs generated more than a billion dollars in sales alone during the same period. And the financial rewards of syndication are astronomical: shows like *The Simpsons* and *The West Wing* did well for their creators in their initial airings on network television, but the real bonanza came from their afterlife as reruns. Syndication has changed the underlying economics of how television shows are conceived and produced, because the rewards of reaching syndication are so much more immense than those generated by the original airing of a show. Every local channel everywhere on the planet that airs an old episode of *Seinfeld* is paying a fee to Jerry Seinfeld, Larry David, and the other creators of the show. Those syndication fees, added up, are mind-boggling: Seinfeld and David together have earned hundreds of millions of dollars from the syndication rights, while earning only a small fraction of that from the show's first run on NBC. Network television made stand-up comics like Milton Berle and Bob Hope millionaires. Syndication has turned today's comics into magnates.

How do the economics of repetition connect to the Sleeper Curve? The virtue of syndication or DVD sales doesn't lie in the financial reward itself, but in the selection criteria that the reward creates in the larger entertainment ecosystem. If the ultimate goal stops being about capturing an audience's attention once, and becomes more about *keeping* their attention through repeat viewings, that shift is bound to have an effect on the content. Television syndication means pretty much one thing: the average fan might easily see a given episode five or ten times, instead of the one or two viewings that you would have expected in the Big Three era. Shows that prosper in syndication do so because they can sustain five viewings without becoming tedious. And sustaining five viewings means adding complexity, not subtracting it. Reruns are generally associated with the dumbing down of popular culture, when, in fact, they're responsible for making the culture smarter. (Syndication has also encouraged another programming trend that has had a neutral impact where the Sleeper Curve is concerned: because viewers often encounter repeat episodes out of sequence—unlike the sequential viewing patterns of a DVD anthology—syndicated episodes that can be viewed in isolation have also prosperered, mostly in the form of the next-generation mystery shows like *Law & Order* and *CSI*. On the whole, the plots of these shows are more intricate than

those of *Dragnet* or *Kojak,* but their insistence on full narrative closure at the end of each episode necessarily puts a
ceiling on their complexity.)

Repetition's impact crater will only deepen in the coming
years. Already, any given episode of a successful television
show will be seen by more people in syndication than it will
during its first run on network TV. As the universe of viewing options expands—inevitably to the point where you can
watch anything in the entire catalogue of television history
anytime you want—the shows that will prosper will be the
ones that can withstand such repeat viewings, while the
more one-dimensional series will grow stale. The success
of *Seinfeld* and *The Simpsons* in syndication—on any given
day, your local cable provider probably pipes a half dozen
episodes of those two shows to your house—demonstrates
that this principle is already at work. In a real sense, this
stands the conventional wisdom about television programming on its head. Aiming for the lowest common denominator might make sense if the show's going to be seen only
once, but with a guarantee of multiple viewings, you can
venture into more challenging, experimental realms and still
be rewarded for it.

To appreciate the magnitude of the shift, you need only
rewind the tape to the late seventies and contemplate the
governing principle that reigned over prime-time programming in the dark ages of *Joanie Loves Chachi*—a philoso-

phy dubbed the theory of "Least Objectionable Program-ming" by NBC executive Paul Klein:

> We exist with a known television audience, and all a show has to be is least objectionable among a segment of the audience. When you put on a show, then, you immediately start with your fair share. You get your 32-share . . . that's about [a third] of the network audience, and the other networks get their 32 shares. We all start equally. Then we can add to that by our competitors' failure—they become objectionable so people turn to us if we're less objection-able. Or, we could lose audience by inserting little "tricks" that cause the loss of audience. . . . Thought, that's tune-out, education, tune-out. Melodrama's good, you know, a little tear here and there, a little morality tale, that's good. Positive. That's least objectionable. It's my job to keep my 32, not to cause any tune-out a priori in terms of ads or concepts, to make sure there's no tune-out in the shows vis-à-vis the competition.

LOP is a pure-breed race-to-the-bottom model: you cre-ate shows designed on the scale of minutes and seconds, with the fear that the slightest challenge—"thought," say, or "education"—will send the audience scurrying to the other networks. Contrast LOP with the model followed by *The Sopranos*—what you might call the Most Repeatable Programming model. MRP shows are designed on the scale of years, not seconds. The most successful programs in the MRP model are the ones you still want to watch three years after they originally aired, even though you've already seen them three times. The MRP model cultivates nuance and depth; it welcomes "tricks" like backward episodes and dense allusions to Hollywood movies. Writing only a few years after Klein's speech, Neil Postman announced that two of television's golden rules were: "Thou shalt have no prerequisites" (meaning that no previous knowledge should be required for viewers to understand a program) and "Thou shalt induce no perplexity." Postman had it right at the time, if you ignored the developing narrative techniques of *Hill Street Blues* and *St. Elsewhere*. But twenty years later, many of the most popular shows in television history regularly flaunt those principles.

The progressive effects of repetition are particularly acute where sales—and not rentals—are concerned. When you're trying to persuade audiences to *purchase* a title, and not simply borrow their attention for thirty minutes, the most successful products are usually the ones that you can imagine watching four years from now, for the fifth time. It's no accident that DVD versions of shows like *The West Wing* and *The Sopranos* have sold more copies than many hit movies. If you're buying a piece of entertainment for your permanent collection, you don't want instant gratification;

you want something that rewards greater scrutiny. The fact that DVD sales now figure so prominently in Hollywood spreadsheets shifts the balance away from films guaranteed to "open" big toward films that cinephiles are likely to add to their permanent collection. (Think of Wes Anderson's films, or Sofia Coppola's, or David Lynch's, or Quentin Tarantino's.) They might lose money at the box office, but they'll turn in a nice profit in DVD sales, and by virtue of their smaller budgets, they don't run the risk of massive failure that wannabe blockbusters do. For the economics of both television and the movie business, the fundamental shift here is from "live" programming to libraries. The studios now mine their libraries of old content for new sales, whether nostalgia DVDs or syndication; and they craft new programming so that it's complex enough to deserve a spot in the home media libraries of consumers. Moving from live to libraries is, ultimately, a shift from Least Objectionable to Most Repeatable.

The success of blisteringly complex narratives like *Memento* and *Eternal Sunshine of the Spotless Mind* showcases the way the MRP model has infiltrated Hollywood. *Eternal Sunshine* screenwriter Charlie Kaufman—who also penned the dizzyingly plotted *Being John Malkovich* and *Adaptation*—described his writing philosophy in an interview on *Charlie Rose,* using language that perfectly contrasts Paul Klein's LOP:

I guess my mindset about movies is that I feel like film is a dead medium. With theater you've got accidents that can happen, performances that can change. But film is a recording. So what I try to do is infuse my screenplays with enough information that upon repeated viewings you can have a different experience. Rather than the movie going linearly to one thing, and at the end telling you what the movie's about—I try to create a conversation with the audience. I guess that's what I try to do— have a conversation with each individual member of the audience.

Kaufman has it exactly right: not just in the sense of rewarding repeat views, but also this idea of creating a "conversation" with the audience. Conversations are two-way affairs; they're participatory by nature. But how do you create a conversation using a "dead medium"? You do it by engaging the minds of the audience, by making them fill in and lean forward. You create plots so complicated and self-referential that you have to work to make sense out of the first viewing—and by the end, all you want to do is rewind the tape and see it over again, just to figure out what you missed.

You can see the Most Repeatable Programming model at work in the narrative transformation of a genre designed explicitly to be viewed dozens of times: children's movies. Because young children have a greater tolerance for repeat encounters with the same story, and because parents of young children have an even greater tolerance for anything that distracts their children long enough for the dishes to be done, the market for DVD and video versions of children's movies is a massive one. Pixar alone has made billions of dollars from the DVD sales of hits such as *Toy Story* and *Monsters, Inc.* This is a market where vast fortunes can be made from content that can sustain ten or twenty viewings (if not more), and so we should expect to see a strong Sleeper Curve driving the complexity and depth of the storytelling as the financial incentives kick in.

And in fact, that's exactly what you find, as we saw in the earlier analysis of children's films over the past few decades. *Finding Nemo* isn't the fastest-selling DVD of all time *in spite* of its complexity; it's the fastest-selling DVD *because* of that complexity. Whenever popular culture shifts its economic incentives from quick hits to long-term repetition, a corresponding increase in quality and depth ensues.

The transformation of video games—from arcade titles designed for a burst of action in a clamorous environment, to contemplative products that reward patience and intense study—provides the most dramatic case study in the power of repetition. The titles that lie at the top of the all-time game best-seller lists are almost exclusively games that can literally be played forever without growing stale: games like *Age of Empires, The Sims,* or *Grand Theft Auto* that have no fixed narrative path, and thus reward repeat play with an ever-changing complexity; sports simulations that allow you to replay entire seasons with new team rosters, or create imaginary leagues with players from different eras. Titles with definitive endings have less value in the gaming economy; the more open-ended and repeatable, the more likely it is that the game will be a breakout hit.

There's a strange antecedent for the Most Repeatable Programming model in the history of moral philosophy: Nietzsche's idea of the "eternal recurrence," his alternative model to Christian morality. Instead of getting people to do the right thing by threatening them with eternal damnation, Nietzsche proposed an alternative structuring myth in which our lives were going to be repeated ad infinitum. If we made a mistake in this life, we'd keep making it forever, which presumably would end up encouraging us not to make mistakes in the first place. Ever since Nietzsche proposed the idea, ethicists and philosophers have been debating its merits as a moral guide, without a clear verdict. But as a governing principle for creating quality pop culture, eternal recurrence makes a lot of sense. Design each title so that it can be watched many times, and you'll end up with more interesting and more challenging culture. And you might just get rich along the way.

*　　*　　*

TECHNOLOGICAL INNOVATION, of course, has contributed mightily to the Sleeper Curve. To begin with, most of the media technologies introduced over the past thirty years have been, in effect, repetition engines: tools designed to let you rewind, replay, repeat. It seems amazing to think of it now, but just thirty years ago, television viewers tuning in for *All in the Family* or *M*A*S*H* had almost no recourse available to them if they wanted to watch a scene again, or catch a bit of dialogue they missed. If you wanted to watch the "Chuckles the Clown" episode of *Mary Tyler Moore* again, you had to wait six months, until CBS reran it during the summer doldrums—and then five years before it started cycling in syndication. The change since then has been so profound that it's hard to remember that television was a pure present-tense medium for half of its existence: what appeared on the screen flew past you, as irretrievable as real-world events. No wonder the networks were so afraid to challenge or confuse; if the show didn't make complete sense the first time around, that was it. There were no second acts.

Since those days, the options for slowing down or reversing time have proliferated: first the VCR, introduced the same year that *Hill Street Blues* appeared; then the explosion of cable channels, running dozens of shows in syndication at any given moment; then DVDs fifteen years later; then TiVo; and now "on demand" cable channels that allow viewers to select programs directly from a menu of options—as well as pause and rewind them. Viewers now curate their own private collections of classic shows, their

DVD cases lining living room shelves like so many triple-decker novels. The supplementary information often packaged with these DVDs adds to their repetition potential: if you're tired of the original episode, you can watch the version with all the deleted scenes spliced in, or listen to a commentary track from the director.

These proliferating new recording technologies are often described as technologies of convenience: you watch what you want to watch, when you want to watch it, as the old TiVo slogan had it. If your *Sopranos*-watching schedule doesn't sync up with the network programmers at HBO, no worries: just order it on demand, or tape it, or TiVo it, or catch it later that week on HBO2. No doubt that convenience is an important selling point, but the technology has another laudable side effect: it facilitates close readings. Fans of *The Sopranos* who want to dissect every scene for subtle references and hidden meanings have half a dozen avenues available to them. Perhaps there would have been fans equally devoted to *Gunsmoke* or *Laverne & Shirley* when those shows first aired, but the technology of that era kept their passions at bay, by limiting the number of times they could watch an episode—which in turn caused the shows' creators to limit the complexity of the programming itself. Instead of adding layers and twists, they went with the least objectionable.

The technological revolutions of the past decade have aided the Sleeper Curve in another way. As technologies of repetition allowed new levels of complexity to flourish, the rise of the Internet gave that complexity a new venue where it could be dissected, critiqued, rehashed, and explained. Years ago I dubbed these burgeoning Web communities "para-sites," online media that latches onto traditional media, and relies on those larger organisms for their livelihood. Public discussion of popular entertainment used to limit itself to the dinner table and the water cooler, but as we saw in the *Apprentice* fan site debate, the meta-conversation has itself grown deeper and more public. Even a modestly popular show—like HBO's critically acclaimed drama *Six Feet Under*—has spawned hundreds of fan sites and discussion forums, where each episode is scrutinized and annotated with an intensity usually reserved for Talmudic scholars. The fan sites create a public display of passion for the show, which nervous Hollywood execs sometimes use to justify renewing a show that might otherwise be canceled due to mediocre ratings. Shows like *Arrested Development* or *Alias* survive for multiple seasons thanks in part to the enthusiasm of their smaller audiences—not to mention the fans' willingness to buy DVD versions en masse when they're eventually released.

These sites function as a kind of decoder ring for the Sleeper Curve's rising complexity. Devoted fans coauthor massive open documents—episode plot summaries, frequently asked questions, guides to series trivia—that exist online as evolving works of popular scholarship, forever being tinkered with by the faithful. Without these new channels, the subtleties of the new culture would be lost to all but the most ardent fans. But the public, collaborative nature of these sites means that dozens or hundreds of fans can team up to capture all the nuances of a show, and leave behind a record for less motivated fans to browse through at their convenience. And so the threshold of complexity rises again. The *Simpsons* creators can bury a dozen subtle film references in each episode and rest assured that their labors will be reliably documented online within a few days. No minor allusion or narrative pirouette will ever go unnoticed, because there are a thousand archivists keeping track at home.

The new possibilities for meta-commentary are best displayed in game walk-throughs: those fantastically detailed descriptions that "walk" the reader "through" the environment of a video game, usually outlining the most effective strategies for completing the game's primary objectives. Hundreds of these documents exist online, almost all of them created by ordinary players, assembling tips and techniques from friends and game discussion boards. They condense the ambiguities and open-ended rule structure of these games into a more linear narrative form—conventionally using a second-person address, as in this walk-though for the game *Half-Life*:

The first task facing you once you make it to the office complex is simply getting down the hallway. About halfway down the hall there's a live wire, randomly discharging electricity into the puddle on the floor. And the door that you can reach is locked. Luckily, there's a ventilation duct just before the live wire. Crawl over to the duct and break the grate with the Crowbar. Be careful, because the discharge can still hit you if you move too far to the right of the grate. Crawl into the duct and follow it to the end. Break the grate and climb into the room. Beware of the Barnacle, and be aware that more will be bursting through the ceiling while you're in the room.

In the corner, you'll see a door with a sign reading "high voltage." Open it, go in, and flip the switch. Now the hallway is safe.

At the other end of the hallway, you'll need to break the window and climb through. The water-filled room to the right has its own electrical problem, but you'll deal with that in a moment. For now, it's time to get some supplies. Go to the left and into the little alcove with the wooden door. . . .

Read a walk-through on its own, without knowing anything about the game it documents, and the text feels like an experimental novel stitched together out of passages stolen from the magazines *Guns & Ammo* and *This Old House*. ("Luckily, there's a ventilation duct just before the live wire. Crawl over to the duct and break the grate with the Crowbar.") For the most part, the stories conveyed by game walk-throughs are unreadable, unless you're in the middle of the game itself, at which point all the stray details and observations carry the force of revelation: "So that's how you get down that hallway!" If you have your doubts about the spatio-logical complexity of today's video games, and don't have the time to sit down and play one yourself, I recommend downloading one of these walk-throughs from the Web and scrolling through it just to gauge the scale and intricacy of these gameworlds.

In the 1930s the Russian mathematician Andrei Kolmogorov arrived at a definition of complexity for any given string of information: the shortest number of bits of information into which the string can be compressed without losing any data. The text string "Smith Smith Smith" is less complex than the string "Smith Jones Bartlett" because you can compress the former into the description "Smith x3." A series of numbers such as "2, 4, 8, 16, 32, 64, etc." is less complex than a random sequence, because you can't express the random sequence with a simple formula. You can think of the text strings of game walk-throughs as compressed versions of the game's original, open-ended state: the walk-throughs document the shortest route from start to finish, with the minimal amount of meandering and false starts. They tell you exactly what you need to know. Judged by the size of these walk-throughs, the Kolmogorov complexity of your average video game has expanded at a prodigious clip. The compressed renditions of *PacMan* came in the form of those famous "patterns": turn left, turn right, turn right again. You could convey the entirety of the *Pac-Man* universe in a few pages of text. By comparison, the walk-through for *Grand Theft Auto III*—by an Australian devotee of the game named Aaron Baker—contains 53,000 words, around the same as the book you are currently reading. Printed out in single-spaced twelve-point type, the document is 164 pages long.

The economics of repetition's race to the top are easy enough to grasp: syndication and DVD sales offer great financial reward to creators who generate titles complex enough to remain interesting through repeat encounters. But where is the economic reward in encouraging meta-commentary? The answer to that puzzle lies in the culture industry's growing emphasis on "thought leaders" or "key influencers." The old way to market a new cultural product was to sell it like detergent: get your brand and your message in front of as many people as possible, and hope to persuade some of them to buy the product. If that means billboards and full-page newspaper ads, great. If that means getting the show in the 8:30 slot after *Cosby,* even better. That's the philosophy of mass marketing, and it may indeed work well for consumer goods where the consumers themselves don't have a huge emotional investment in the product. But where culture is concerned—movies, books, television shows—people don't just build relationships with products based on the dictates of mass advertising. Word of mouth is often more powerful, and where word of mouth is concerned, some consumers speak louder than others. They're the early adopters; the ones who pride themselves on their pop culture mastery, their eye for new shows and rising talent.

The meta-commentary sites have endowed these arm-chair experts with venues where their expertise can flourish in public. Before the Internet, a rabid fan who wanted to compose a 53,000-word inventory of his favorite video game didn't have an easy way to get his opus in the hands of people who might be interested in reading it—short of distributing xeroxed copies on the sidewalk. Now the experts can convey their wisdom to tens of thousands of eager recipients desperately trying to reach the second city in *Grand Theft Auto* or figure out why Tony Soprano had that guy killed last night. There's no real financial reward for these key influencers and mavens themselves; Aaron Baker doesn't write 164-page walk-throughs because he thinks they'll make him rich. He does it for the public pride he takes in creating the authoritative guide to one of the most popular games of all time. (There are social rewards, in other words, not financial ones.) But a significant financial reward does exist for entertainment creators who attract people like Aaron Baker to their products, because it is precisely those experts who end up persuading other people to watch the show or play the game or see the movie. The way to attract the Aaron Bakers of the world is to make products complex enough that they *need* experts to decipher them. Key influencers like to think of themselves as operating on the cutting edge, detecting patterns or trends in cultural forms that ordinary consumers don't perceive until someone points them out. The way to attract these experts, then, is to give them material that challenges their decoding skills, material that lets them show off their chops. Instead of rewarding the least offensive programming, the system rewards the titles that push at the edges of convention, the titles that welcome close readings. You can't win over the aficionados with the lowest common denominator.

*   *   *

TECHNOLOGY AMPLIFIES the Sleeper Curve in one final respect: it introduces new platforms and genres at an accelerating rate. We had thirty years to adapt to the new storytelling possibilities of cinema; then another twenty for radio; then twenty years of present-tense television. And then the curve slants upward: five years to acclimate to the VCR and video games; then e-mail, online chats, DVDs, TiVo, the Web—all becoming staples of the pop culture diet in the space of a decade. McLuhan had a wonderful term for this accelerating sequence, "electric speed":

Today it is the instant speed of electric information that, for the first time, permits easy recognition of the patterns and the formal contours of change and development. The entire world, past and present, now reveals itself to us like a growing plant in an enormously accelerated movie. Electric speed is synonymous with light and with the understanding of causes.

McLuhan believed that this rate of change shed light on the hitherto invisible ways in which media shaped a given society's worldview; it let us see the impact of the medium, and not just the message. When your culture revolves exclusively around books for hundreds of years, you can't detect the subtle ways in which the typographic universe alters your assumptions. But if you switch from cinema to radio to television in the course of a lifetime, the effects of the different media become apparent to you, because you have something to measure them against. That enlightenment is a profound thing, but it is only part of the legacy of electric speed. Adapting to an ever-accelerating sequence of new technologies also trains the mind to explore and master complex systems. When we marvel at the technological savvy of average ten-year-olds, what we should be celebrating is not their mastery of a specific platform—Windows XP, say, or the GameBoy—but rather their seemingly effortless ability to pick up new platforms on the fly, without so much as a glimpse at a manual. What they've learned is not just the specific rules intrinsic to a particular system; they've learned abstract principles that can be applied when approaching *any* complicated system. They don't know how to program a VCR because they've memorized the instructions for every model on the market; they know how to program a VCR because they've learned general rules for probing and exploring a piece of technology, rules that come in handy no matter what model VCR you put in front of them.

Cognitive scientists have argued that the most effective learning takes place at the outer edges of a student's competence: building on knowledge that the student has already acquired, but challenging him with new problems to solve. Make the learning environment too easy, or too hard, and students get bored or frustrated and lose interest. But if the environment tracks along in sync with the students' growing abilities, they'll stay focused and engaged. The game scholar James Paul Gee has observed precisely this phenomenon—called the "regime of competence" principle—at work in the architecture of successful video games. "Each level dances around the outer limits of the player's abilities," he writes, "seeking at every point to be hard enough to be just doable . . . which results in a feeling of simultaneous pleasure and frustration—a sensation as familiar to gamers as sore thumbs." Game designers don't build learning machines out of charity, of course; they do it because there's an economic reward in creating games that stay close to that border. Make a game too hard, and no one will buy it. Make it too easy, and no one will buy it. Make a game where the challenges evolve alongside your skills, and you'll have a shot at success. And you'll have built a powerful educational tool to boot.

I think the regime of competence principle operates on another scale as well: not in the forty hours it takes to complete your average video game, but on the hundred-year scale of electric speed. When cinema first became a mainstream diversion in the early 1900s, the minds of that era were not primed to master ten new technologies and dozens of new genres in the next decade; they had to adapt to the new conventions of moviegoing, learning a new visual language, and a new kind of narrative engine. But as the new technologies started to roll out in shorter and shorter cycles, we grew more comfortable with the process of probing a new form of media, learning its idiosyncrasies and its distortions, its symbolic architecture and its rules of engagement. The mind adapts to adaptation. Eventually you get a generation that welcomes the challenge of new technologies, that embraces new genres with a flexibility that would have astonished the semi-panicked audiences that trembled through the first black-and-white films.

Technology manufacturers have an economic incentive to obey the regime of competence principle as well: if your new platform—an operating system, say, or a wireless communicator, or TiVo-style personal video recorder—is too familiar, it will seem like old news to potential consumers; but if you push too far past the regime of competence, you'll lose your audience as well. Release new technologies that challenge the mind without overtaxing it, and release them in shorter and shorter cycles, and the line that tracks our abilities to probe and master complex systems will steadily ascend, turning upward in a parabolic climb as the cycles of electric speed increase.

Project that data over a hundred years, and you will have a chart that looks remarkably like the Flynn Effect.

*   *   *

POP CULTURE'S race to the top over the past decades forces us to rethink our assumptions about the base tendencies of mass society: the *Brave New World* scenario, where we're fed a series of stupefying narcotics by media conglomerates interested solely in their lavish profits with no concern for the mental improvement of their consumers. As we've seen, the Sleeper Curve isn't the result of media titans doing charitable work; there's an economic incentive in producing more challenging culture, thanks to the technologies of repetition and meta-commentary. But the end result is the same: left to its own devices, following its own profit motives, the media ecosystem has been churning out popular culture that has grown steadily more complex over time. Imagine a version of *Brave New World* where *soma* and the *feelies* make you smarter, and you get the idea.

If the Sleeper Curve turns the conventional wisdom about

mass culture on its head, it does something comparable to our own heads—and the truisms we like to spread about them. Almost every Chicken Little story about the declining standards of pop culture contains a buried blame-the-victim message: Junk culture thrives because people are naturally drawn to simple, childish pleasures. Children zone out in front of their TV shows or their video games because the mind seeks out mindlessness. This is the Slacker theory of brain function: the human brain desires above all else that the external world refrain from making it do too much work. Given their druthers, our brains would prefer to luxuriate among idle fantasies and mild amusements. And so, never being one to refuse a base appetite, the culture industry obliges. The result is a society where maturity, in Andrew Solomon's words, is a "process of mental atrophy."

These are common enough sentiments, but they contain a bizarre set of assumptions if you think about them from a distance. Set aside for the time being the historical question of why IQs are climbing at an accelerating rate while half the population wastes away in mental atrophy. Start instead with the more basic question of why our brains would actively seek out atrophy in the first place.

The *Brave New World* critics like to talk a big game about the evils of media conglomerates, but their worldview also contains a strikingly pessimistic vision of the human mind. I think that dark assumption about our innate cravings for junk culture has it exactly backward. We know from neuroscience that the brain has dedicated systems that respond to—and seek out—new challenges and experiences. We are a problem-solving species, and when we confront situations where information needs to be filled in, or where a puzzle needs to be untangled, our minds compulsively ruminate on the problem until we've figured it out. When we encounter novel circumstances, when our environment changes in a surprising way, our brains lock in on the change and try to put it in context or decipher its underlying logic.

Parents can sometimes be appalled at the hypnotic effect that television has on toddlers; they see their otherwise vibrant and active children gazing silently, mouth agape at the screen, and they assume the worst: the television is turning their child into a zombie. The same feeling arrives a few years later, when they see their grade-schoolers navigating through a video game world, oblivious to the reality that surrounds them. But these expressions are not signs of mental atrophy. They're signs of *focus*. The toddler's brain is constantly scouring the world for novel stimuli, precisely because exploring and understanding new things and experiences is what learning is all about. In a house where most of the objects haven't moved since yesterday, and no new people have appeared on the scene, the puppet show on the television screen is the most surprising thing in the child's environment, the stimuli most in need of scrutiny and explanation. And so the child locks in. If you suddenly plunked down a real puppet show in the middle of the living room, no doubt the child would prefer to make sense of that. But in most ordinary household environments, the stimuli onscreen offer the most diversity and surprise. The child's brain locks into those images for good reason.

Think about it this way: if our brain really desired to atrophy in front of mindless entertainment, then the story of the last thirty years of video games—from *Pong* to *The Sims*—would be a story of games that grew increasingly simple over time. You'd never need a guidebook or a walk-through; you'd just fly through the world, a demigod untroubled by challenge and complexity. Game designers would furiously compete to come out with the simplest titles; every virtual space would usher you to the path of least resistance. Of course, exactly the opposite has occurred. The games have gotten more challenging at an astounding rate: from *PacMan*'s single page of patterns to *Grand Theft Auto III*'s 53,000-word walk-through in a mere two decades. The games are growing more challenging because there's an economic incentive to make them more challenging—and that economic incentive exists because our brains *like* to be challenged.

If our mental appetites draw us toward more complexity and not less, why do so many studies show that we're reading fewer books than we used to? Even if we accept the premise that television and games can offer genuine cognitive challenges, surely we have to admit that books challenge different, but equally important, faculties of the mind. And yet we're drifting away from the printed page at a steady rate. Isn't that a sign of our brains gravitating to lesser forms?

I believe the answer is no, for two related reasons. First, most studies of reading ignore the huge explosion of reading (not to mention writing) that has happened thanks to the rise of the Internet. Millions of people spend much of their day staring at words on a screen: browsing the Web, reading e-mail, chatting with friends, posting a new entry to one of those 8 million blogs. E-mail conversations or Web-based analyses of *The Apprentice* are not the same as literary novels, of course, but they are equally text-driven. While they suffer from a lack of narrative depth compared to novels, many online interactions do have the benefit of being genuinely two-way conversations: you're putting words together yourself, and not just digesting someone else's. Part of the compensation for reading less is the fact that we're writing more.

The fact that we are spending so much time online gets to the other, more crucial, objection: yes, we're spending less time reading literary fiction, but that's because we're spending less time doing *everything* we used to do before. In fact, the downward trend that strikes the most fear in the hearts of Madison Avenue and their clients is not the decline of literary reading—it's the decline of television watching. The most highly sought demographic in the country— twenty-something males—watches almost one-fifth less television than they did only five years ago. We're buying fewer CDs; we're going out to the movies less regularly. We're doing all these old activities less because about a dozen new activities have become bona fide mainstream pursuits in the past ten years: the Web, e-mail, games, DVDs, cable on-demand, text chat. We're reading less because there are only so many hours in the day, and we have all these new options to digest and explore. If reading were the only cultural pursuit to show declining numbers, there might be cause for alarm. But that decline is shared by all the old media forms across the board. As long as reading books remains *part* of our cultural diet, and as long as the new popular forms continue to offer their own cognitive rewards, we're not likely to descend into a culture of mental atrophy anytime soon.

*   *   *

Now for the bad news. The story of the last thirty years of popular culture is the story of rising complexity and increased cognitive demands, an ascent that runs nicely parallel to—and may well explain—the upward track of our IQ scores. But there are hidden costs to the Sleeper Curve. It's crucial that we abandon the *Brave New World* scenario where mindless amusement always wins out over more challenging fare, that we do away once and for all with George Will's vision of an "increasingly infantilized society." Pop culture is not a race to the bottom, and it's high time we accepted—even celebrated—that fact. But even the most salutary social development comes with peripheral effects that are less desirable.

The rise of the Internet has forestalled the death of the typographic universe—and its replacement by the society of the image—predicted by McLuhan and Postman. Thanks to e-mail and the Web, we're reading text as much as ever, and we're writing more. But it is true that a specific, historically crucial kind of reading has grown less common in this society: sitting down with a three-hundred-page book and following its argument or narrative without a great deal of distraction. We deal with text now in shorter bursts, following links across the Web, or sifting through a dozen e-mail messages. The breadth of information is wider in this world, and it is far more participatory. But there are certain types of experiences that cannot be readily conveyed in this more connective, abbreviated form. Complicated, sequential works of persuasion, where each premise builds on the previous one, and where an idea can take an entire chapter to develop, are not well suited to life on the computer screen. (Much less life on *The O'Reilly Factor*.) I can't imagine getting along without e-mail, and I derive great intellectual nourishment from posting to my weblog, but I would never attempt to convey the argument of this book in either of those forms. Postman gets it right:

To engage the written word means to follow a line of thought, which requires considerable powers of classifying, inference-making and reasoning. . . . In the eighteenth and nineteenth centuries, print put forward a definition of intelligence that gave priority to the objective, rational use of the mind and at the same time encouraged forms of public discourse with series, logically ordered content. It is no accident that the Age of Reason was coexistent with the growth of a print culture, first in Europe and then in America.

Networked text has its own intellectual riches, of course: riffs, annotations, conversations—they all flourish in that ecosystem, and they all can be dazzlingly intelligent. But they nonetheless possess a *different kind* of intelligence from the intelligence delivered by reading a sustained argument for two hundred pages. You can convey attitudes and connections in the online world with ease; you can brainstorm with twenty strangers in a way that would have been unthinkable just ten years ago. But it is harder to transmit a fully fledged worldview. When you visit someone's weblog, you get a wonderful—and sometimes wonderfully intimate—sense of their voice. But when you immerse yourself in a book, you get a different sort of experience: you enter the author's mind, and peer out at the world through their eyes.

Something comparable happens in reading fiction as well.

No cultural form in history has rivaled the novel's capacity to re-create the mental landscape of another consciousness, to project you into the first-person experience of other human beings. Movies and theater can make you feel as though you're part of the action, but the novel gives you an inner vista that is unparalleled: you are granted access not just to the events of another human's life, but to the precise way those events settle in his or her consciousness. (This is most true of the modernist classics: James, Eliot, Woolf, Conrad.) Reading *Portrait of a Lady*—once you've shed your MTV-era expectations about pacing and oriented yourself to James's byzantine syntax—you experience another person thinking and sensing with a clarity that can be almost uncanny. But that cognitive immersion requires a physical immersion for the effect to work: you have to commit to the book, spend long periods of time devoted to it. If you read only in short bites, the effect fades, like a moving image dissolving into a sequence of frozen pictures.

So the Sleeper Curve suggests that the popular culture is not doing as good a job at training our minds to follow a sustained textual argument or narrative that doesn't involve genuine interactivity. (As we've seen in gaming culture, kids are incredibly talented at focusing for long stretches when the form is truly participatory.) The good news, of course, is that kids aren't being exclusively educated by their Nintendo machines or their cell phones. We still have schools and parents to teach wisdom that the popular culture fails to impart. The Dr. Spock manual had it half right after all: parents should "foster in [their] children a love of reading and the printed word from the start." They just shouldn't underestimate the virtues of other media as well.

But what about all the sex and violence? Having made the case for the cognitive challenges of today's popular culture, it's only fair to return to the question of morals. Even if you accept the premise that a whole host of intellectual tools— our pattern recognition skills, our ability to probe and telescope, to map complicated narratives—have been enhanced by progressive trends in the popular culture, you can still reasonably object that all those improvements don't cancel out the declining moral or behavioral standards advocated by these forms. In which case the Sleeper Curve would only be a consolation prize—we're raising a generation of cognitive superstars who are nonetheless ethically rudderless. Intelligent, yes, but without values.

I question that scenario for several reasons. First, I suspect we seriously overestimate the extent to which our core values are transmitted to us via the media. Most people understand that the characters on the screen are fictitious ones, and their flaws are there to amuse and entertain us, and not give us ethical guidance. Parents and peer groups are still vastly more influential where values are concerned than Tony Soprano or the carjackers of *Grand Theft Auto*. And the truth is most shows and games and movies still gravitate toward traditional morality play structures in the end: the good guys still win out, and they usually do it by being honest and playing by the rules. For every *Sopranos* or *Grand Theft Auto* there are a dozen *West Wing*s or *Zelda*s, fairy tales of earnest good intentions and civic pride.

That some of the culture today does push at the boundaries of acceptable or healthy moral values shouldn't surprise us, because it is in the nature of myth and storytelling to explore the edges of a society's accepted beliefs and conventions. Popular stories rarely flourish in environments of perfect moral clarity; they tend to blossom at exactly the spaces where some established order is being questioned or tested. We're still retelling the Oedipus myth precisely because it revolved around the violation of fundamental human values. Stories of perfectly happy families—where all laws are obeyed and no values are challenged—don't captivate us in the same way. (Even *The Brady Bunch* required two preexisting nuclear families to break up for its own narrative to take flight.) So when we see the popular culture exploring behavior that many see as morally bankrupt, we need to remind ourselves that deviating from an ethical norm is not just an old story. In a real sense, it's where stories begin.

Certainly it is true that the media today is more violent than it has ever been before, at least in terms of the physical carnage willingly re-created on the screen. Violence has always been a constant in the narratives we tell ourselves—it's part of that tendency of narrative to seek out the extremes of human experience. The difference now is that we get to see the bodily details of that violence in ways that would have been unimaginable just fifty years ago. Video games, in particular, have grown dramatically more violent in the past fifteen years, as the graphical capabilities of the modern PC have enabled ever more realistic displays of bloodshed.

The question is whether that violence has an effect on the mind that apprehends it. It should go without saying at this point that I believe different forms of media can alter our brains in significant ways; the premise of the Sleeper Curve adheres to that principle: more complex popular entertainment is creating minds that are more adept at certain kinds of problem-solving. But violence is part of the *content* of popular media, and as I have explained throughout the preceding pages, the content of most entertainment has less of an impact than the kind of thinking the entertainment forces you to do. This is why we urge parents to instill a general love of reading in their children, without worrying as much about *what* they're reading—because we believe there is a laudable cognitive benefit that comes just from the act of reading alone, irrespective of the content. The same principle applies to television or film or games.

By any measure, the content of a *24* episode is more violent and disturbing than an episode of *My Three Sons*. But *24* makes the viewer *think* in ways that earlier shows never dared; it makes them analyze complex situations, track social networks, fill in information withheld by the creators. The great majority of television viewers understand that the violence they encounter on these contemporary shows is fiction; they understand that they should not look to Tony Soprano for moral guidance, or model their real-world driving on their *Grand Theft Auto* excursions. But the mental exercise they undergo in watching these shows or playing these games is not fiction. Think of those test subjects whose visual intelligence improved after playing the war game *Medal of Honor*; they trained their perceptual systems to perform at a higher level by running around shooting at things in a military simulation. That much is clear. The question is whether that experience also made them more likely to pick up a gun in actual life, more likely to resort to violence in solving real-world problems.

If the subject matter of popular entertainment truly had a significant impact on our behavior (and especially the behavior of the younger generations) then logically we should expect to see very different trends in real-world society. Over the last ten years—a period of unprecedented *fictional* violence in the American household, thanks to *Quake* and Quentin Tarantino films and Tony Soprano—the country simultaneously experienced the most dramatic drop in violent crime in its history. Yes, the Columbine shooters were most likely influenced by playing violent games like *Quake,* but as tragic as that event was, we don't analyze social trends by looking at isolated single examples; we look at broad patterns in the society, and the broad pattern of the last decade is less violence, not more. That improvement is most telling among precisely the demographic groups allegedly at risk for media-influenced violence. In late 2004, the Departments of Justice and Education released a joint study that showed violent crime in the nation's schools had been literally cut in half over the ten-year period from 1992 to 2002, dropping from forty-eight to twenty-four incidents per 100,000 students.

Now, it is theoretically possible that violent media has nevertheless been provoking violent acts throughout that period, but those effects have been masked by the other, pacifying forces at work in society: better policing, higher incarceration rates, or low unemployment. Perhaps we would have had only *ten* violent acts per 100,000 students if it weren't for *Grand Theft Auto*. (Of course, it's just as likely that exposure to violent media—particularly in the participatory mode offered by games—functions as a safety valve for kids who might otherwise be inclined to express their aggression in the real world, and thereby causes violence to decrease.) The one thing we know for certain is this: If there is some positive correlation between exposure to fictional violence and violent behavior, its effects are by definition much weaker than the other social trends that shape violence in society.

Does that mean anything goes? I'm often asked what the

Sleeper Curve means for the practical decisions that parents have to make about regulating their children's spare time. I realize that, in writing this book, I have set myself up to be misrepresented as the guy who argues that kids should be allowed to play *Doom* all day, and never open a novel. So let me be clear for the parents who are reading this. Yes, the trends are toward more media complexity; yes, games and television shows and films have cognitive rewards that we should better understand and value. But some of those cultural works are more rewarding than others.

In pointing out some of the ways that popular culture has improved our minds, I am not arguing that parents and other caregivers should stop paying attention to the way their children amuse themselves. What I *am* arguing for is a change in the criteria we use to determine what really is cognitive junk food, and what is genuinely nourishing. Instead of worrying about a show's violent or tawdry content, instead of agitating over wardrobe malfunctions or the f-word, the true test should be whether a given show engages or sedates the mind. Is it Least Objectionable Programming, or Most Repeatable Programming? Is it a single thread strung together with predictable punch lines every thirty seconds? Or does it map a complex social network? Is your onscreen character running around shooting everything in sight, or is she trying to solve problems and manage resources? If your kids want to watch reality TV, encourage them to watch *Survivor* over *Fear Factor.* If they want to watch a mystery show, encourage *24* over *Law & Order.* If they want to play a violent game, then encourage *Grand Theft Auto* over *Quake.* (Indeed, it might be just as helpful to have a rating system that uses mental labor and not obscenity and violence as its classification scheme for the world of mass culture.) For parents, if your selection principle is built around cognitive challenge, and not content, then you needn't limit your children's media intake to dutiful nightly exposure to Jim Lehrer and *NOVA*; the popular culture is supplying plenty of vigorous cognitive workouts on its own.

Where our media diets are concerned for all of us—young, old, or somewhere in the middle—the commonsense rule still applies: moderation in everything. However laudable *SimCity* is, if you've spent the last week locked in your study playing it, you should pick up a book for a change. (And preferably not a *SimCity* game guide.) But neither should we deny ourselves the occasional obsession. These are deep, rich worlds being created on our screens; you can't truly experience them—you can't probe their physics and telescope your way through their multiple objectives—without becoming a little obsessed in the process. Out of obsession comes expertise, a confidence in your own powers of analysis—a sense that if you stick with the system long enough, you'll truly figure out how it works.

Kids and grownups both can learn from those obsessions. In fact, one of the unique opportunities of this cultural moment lies precisely in the blurring of lines between kid and grownup culture: fifty-year-olds are devouring Harry Potter; the median age of the video game–playing audience is twenty-nine; meanwhile, the grade-schoolers are holding down two virtual jobs to make ends meet with a virtual family of six in *The Sims*. Most of the defining popular diversions of our time—Pixar movies, *The Lord of the Rings*, *Survivor*—possess genuine appeal for ten-year-olds, GenXers, and boomers alike. Writing in *The New Yorker* a few years ago, the writer Kurt Andersen adroitly described this trend:

> More than any other person, Steven Spielberg is responsible for this magnificent demographic blur. He invented the signal modern Hollywood hybrid—high-end Saturday matinées for grownups, children's movies that adults unashamedly want to see, like "Indiana Jones" and "Jurassic Park." . . . Our parents may have glanced at "The Flintstones," but it was no grownup's favorite show; "The Simpsons" and "King of the Hill" and "South Park" are.

Too often we imagine the blurring of kid and grownup culture as a series of violations: the nine-year-olds who have to have nipple rings explained to them thanks to Janet Jackson; the suburban teenagers reciting gangsta rap lyrics instead of the Pledge of Allegiance. But this demographic blur has a commendable side that we don't acknowledge enough. The kids are forced to think like grownups: analyzing complex social networks, managing resources, tracking subtle narrative intertwinings, recognizing long-term patterns. The grownups, in turn, get to learn from the kids: decoding each new technological wave, parsing the interfaces, and discovering the intellectual rewards of play. Parents should see this as an opportunity, not a crisis. Smart culture is no longer something you force your kids to ingest, like green vegetables. It's something you share.

*   *   *

I HAVE ALMOST no record of the dice-baseball games that I designed myself all those years ago: only a fragment of player cards from the '79 Yankees. But thanks to the infinite storage of eBay, I now have some of my favorite games from that stage of my life sitting beside me in my study: APBA, Strat-o-Matic, even Extra Innings. Every now and then I'll pull one of them out and flip through the player cards and charts. The encounter never fails to leave me in a strange sort of reverie. On the one hand, the colors and shapes—even the typefaces—of the games are all wonderfully familiar. But at the same time, a powerful distance has opened up between these games and my adult self. I once spent one entire evening scouring the Extra Innings binder, with its endless rows of data, trying to marshal all my intellectual powers to figure out how the game was actually played. I could have ploughed through the instructions, of course, but I wanted to do it the hard way, because I had once known the rules of this game as intimately as anything in my life—and besides, I was only ten years old at the time! How hard could it be? But the longer I looked at the charts, the more the game seemed like a cipher to me, like some kind of numerical programming language that I had never learned. And with that mystery came a kind of wonder: not that my ten-year-old self had been capable of learning this language—kids are capable of amazing feats of cognition, after all—but that I had possessed the dedication and stamina to master such a complex system, without anyone actually forcing me to learn it.

When I think back to my ten-year-old self, sprawled on my bedroom floor, consulting my dice-baseball charts as though they were some kind of statistical scripture, I can see all the defining characteristics of the Sleeper Curve lurking there, in embryo. I was amusing myself, no doubt, but the amusement came from the challenge of probing a virtual world, learning and inventing its rules along the way. Each game that arrived in the mail, each game that I designed myself, offered an intoxicating new universe to explore. Eventually, I found that I liked the process of picking up a new game more than I liked actually playing them. There were no interesting narratives that emerged out of my dice-baseball obsessions, and no moral instruction. I suspect my people skills suffered somewhat for all those hours locked alone in my room. But I am convinced that during this phase of my life, no other activity—in the classroom or anywhere else—engaged my mind with as much focus and conceptual rigor. I was learning how to *think* there on the floor with my twenty-sided dice and my situation charts. It might not have looked like much—but then again, neither does sitting around with your nose in a book.

Those years I passed with my baseball simulations are now a routine rite of passage for most kids today, whether they're probing the worlds of *Zelda,* or learning new communication protocols, or tracking the multiple threads of *Finding Nemo*. Believing in the Sleeper Curve does not mean that teachers or parents or role models have become obsolete. It does not mean that we should give up on reading and let our kids spend all their free time tethered to the Xbox. But it does mean that we should discard, once and for all, a number of easy assumptions we like to make about the state of modern society. The cultural race to the bottom is a myth; we do not live in a fallen state of cheap pleasures that pale beside the intellectual riches of yesterday. And we are not innate slackers, drawn inexorably to the least offensive and least complicated entertainment available. All around us the world of mass entertainment grows more demanding and sophisticated, and our brains happily gravitate to that newfound complexity. And by gravitating, they make the effect more pronounced. Dumbing down is not the natural state of popular culture over time—quite the opposite. The great unsung story of our culture today is how many welcome trends are going up.

NOTES ON
FURTHER READING

Games

IF YOU DON'T COUNT the game guides, the body of
work assessing video game culture is surprisingly thin, given
how massive the gaming industry has become. But a few
thoughtful texts exist, starting with J. C. Herz's pioneering
Joystick Nation. Steven Poole's *Trigger Happy* and sections
of Douglas Rushkoff's *Playing the Future* feature insightful
analysis of gaming culture. The scholar James Paul Gee has
done the most interesting work on the cognitive effects of
gameplay—particularly in his book *What Video Games
Have to Teach Us About Learning and Literacy*. Many
fascinating experiments in using games as educational

tools have come out of the Education Arcade consortium (educationarcade.org), whose cofounder Henry Jenkins has been the model of the pop culture public intellectual, making a number of crucial defenses of games in the media and in the courtroom. Some of the ideas presented here about the logic of gaming are explored from a game designer's point of view in *Rules of Play*, a textbook coauthored by the designer Eric Zimmerman. The field of video game theory is sometimes called "ludology"; for further reading about this nascent critical movement, I recommend the Web sites ludology.org and seriousgames.org. Readers interested in the way gaming culture is transforming business will want to check out two relatively new books: *Got Game*, by John Beck and Mitchell Wade, and Pat Kane's delightful manifesto *The Play Ethic*.

Culture-as-System

IN THE INTRODUCTION, I explained that my approach in this book would be more systemic than symbolic, analyzing the forces that bring about a certain cultural form, and not decoding its meaning. I do not want to be misinterpreted here: clearly cultural works do have a direct symbolic relationship to their sociohistorical context, and there are situations where explicating those symbolic relationships can be a productive enterprise. A symbolic or representational intepretation lends itself most directly to what we used to call without scare quotes the Great Books, as opposed to middlebrow popular culture. The classics—and the soon-to-be classics—are in their own right descriptions and explanations of the cultural systems that produced them. *Middlemarch* is both a good story *and* an analysis of mid-nineteenth-century British culture. You could write a book—in fact, many have been written—on how *Middlemarch* represents the challenges and complexities of that culture. But in doing so you're creating a work of *appreciation* and not explanation. The question you're asking is: "What is George Eliot trying to say here?" The questions raised in this book, on the other hand, take a different form. The question is not: "What are the creators of *Grand Theft Auto* trying to say?" The question is: "How did *Grand Theft Auto* come to exist in the first place? And what effects does it have on the people who play it?"

And even that formulation is too specific, because it's not *Grand Theft Auto* that we're ultimately interested in explaining; it's the general cultural tendencies of which *Grand Theft Auto* is a representative example. This is a crucial way in which mass culture differs from high art: with mass culture, the individual works are less interesting than the broader trends, and the interesting question to ask of those trends is where they come from, what kind of cultural ecosystem encourages their development. The advantage of this systemic approach is that it gets you out of the

"Madonna Scholars" syndrome. The talk-show hosts and conservative commentators love to poke fun at academics studying lowbrow culture, precisely because they assume that these scholars have the audacity to study "Like a Virgin" in the same way that they would dissect *Remembrance of Things Past*. But if you're looking at the work as part of a larger set of cultural trends, and looking at different scales of experience, then the critique doesn't stick, because what you're ultimately interested in is the way culture affects human minds, not the sanctity of the individual work of art. And right now, like it or not, Madonna has more mindshare than Proust does. (Even if she hasn't had a hit album in a few years.)

This systemic approach, while still not exactly mainstream, has grown increasingly common over the past few years, in both academic and popular forms of commentary. The philosophical attack on symbolic criticism begins in many ways with Gilles Deleuze and Felix Guattari's revolutionary treatises *Anti-Oedipus* and *A Thousand Plateaus*—two almost impossibly dense and allusive works that dismantled the then dominant structure of signifier/signified, replacing it with a complex system of multiple interacting flows. Instead of allegorical trees, Deleuze and Guattari proposed a "rhizome" network model that borrowed extensively from the language of complexity theory. The Deleuzian model grew more useful in the hands of the brilliant and eclectic Manuel De Landa, whose writing analyzed the development of medieval towns, the patterns of language evolution, and the history of weapons all through the lens of complex systems theory. (His book *A Thousand Years of Nonlinear History* is a mind-bending read.)

The fashionable notion of "memes"—originally coined by Richard Dawkins almost as an afterthought in his 1976 book *The Selfish Gene*—also takes a systems approach to the history of culture: like genes themselves, successful ideas (or memes) thrive because they're good at reproducing themselves in other minds, and thus spreading through the population. Their symbolic fitness—their ability to represent or describe the world—is only a secondary value; the defining attribute of the meme is not whether or not it's true, but whether it is capable of reproducing itself, and whether it belongs to a wider system of memes (sometimes called a memeplex) that foster its replication. I recommend Susan Blackmore's artful and eloquent *The Meme Machine* as an introduction to the emerging science of memetics. Though he emphasizes the interpersonal connections that direct the flow of ideas, Malcolm Gladwell's best-selling *The Tipping Point* made a comparable argument using the language of epidemics. Some cultural trends happen, Gladwell argued, because of feedback loops that have little to do with the content of the trend itself: a wave of interest in Hush Puppies surges through society not because the fifties iconography of the shoe represents a desire to return to the simpler values of that earlier time, but rather because the complex system of fashion is filled with threshold points where some new trend starts a self-reinforcing cycle that propels it into national popularity. The shoe is no more an allegory than a brutal flu season is. Douglas Rushkoff had used similar contagion metaphors in his 1993 book *Media Virus,* and while his later *Playing the Future* relied on more symbolic and zeitgeist criticism, it remains probably the closest book in spirit to the argument I have laid out here.

Consilience

APPROACHING POPULAR CULTURE as a complex system of interacting forces necessitates traversing different scales of experience in your analysis. This level-jumping should be familiar from the preceding pages: we looked at the evolution of the storytelling engines of TV dramas as though we were narratologists; the discussion of the rise of meta-commentary might have belonged to a McLuhan-style analysis of new media; the exploration of the brain's reward architecture drew heavily from the latest in neuroscience. The movement from discipline to discipline can't be a simple case of intellectual tourism; the different scales must *connect* to each other, in a kind of consilient chain. The narratological approach explains what's new in the formal structure of a television series or video game; the economic and technological analysis explains the conditions that made that structure possible; and the neuroscience explains why people find the structure appealing in the first place. Each level produces information that is in turn passed down to the next level for analysis.

A map of that chain would look something like this:

Narratology/Semiotics	⟶ The Work
Media Theory	⟶ The Platform
Economics	⟶ The Market
Sociology	⟶ The Audience
Neuroscience	⟶ The Mind

Each level produces a series of questions that can only be answered by a level further down the chain. Leave one of those levels out, and the overall picture suffers; blind spots appear in the argument. Focus exclusively on one level and ignore all the others, and the whole interpretative act shifts from explanation to description. You have to climb the entire ladder to get the story right.

One rung on that ladder stands out: the neuroscience. Cultural criticism has a long history of ignoring the sciences (hard and soft), and a recent history of outright hostility in the many attempts to deconstruct or relativize the "truth claims" of science. I think of the so-called science wars as a tremendous wasted opportunity: antagonizing both sides of the divide, and blinding both sides to the many productive compatibilities that do exist. In fact, if you tune out much of that bombast, there's quite a bit in the structuralist and post-structuralist tradition that dovetails with new developments in the sciences. To give just a few examples: The underlying premise of deconstruction—that our systems of thought are fundamentally shaped and limited by the structure of language—resonates with many chapters of a book like Steven Pinker's *The Language Instinct*, despite the fact that Pinker himself has launched a number of attacks on recent cultural theory. The postmodern assumption of a "constructed reality" goes nicely with the idea of consciousness as a kind of artificial theater and not a direct apprehension of things in themselves. Semiotics and structuralism both have roots in Levi-Strauss's research into universal mythology, which obviously has deep connections to the project of evolutionary psychology. And De Landa has amply demonstrated the fundamental alliance between Deleuzian philosophy and complexity theory, an alliance that goes back to Deleuze's interest in the work of Nobel laureate (and founding complexity theorist) Ilya Prigogine.

And so in climbing the ladder of consilience, we can't afford to draw an arbitrary line at the sciences; too many productive connections exist. If McLuhan is right and media are extensions of our central nervous system, then we need a theory of the central nervous system as much as we need a theory of media; if the network technology we're creating takes the form of self-organizing systems, then we need the tools of complexity theory to make sense of those networks. But neither should we grant the sciences a de facto supremacy over the other levels in the interpretative model. In this book's argument, neuroscience arrives at several key points to explain the interaction between media and mind, but it's certainly not correct to describe my argument as ultimately reducing everything down to the firing of neurons. When you're trying to tell the story of how a hurricane came to do $50 billion worth of damage, the economic story of barrier island real estate development is just as important as the story of oceanic currents. The same goes for the story of how video games came to sharpen our minds: you need intelligence testing and narrative theory and brain imaging and economics to tell that story accurately, and none of those elements holds a trump card over the others.

It seems to me that the dialogue between the humanities and the sciences has been steadily growing in civility—and fruitful exchange—over the past decade. To my mind the most interesting work right now is work that tries to bridge the two worlds, that looks for connections rather than divisions. This is ultimately what E. O. Wilson was proposing in *Consilience*: not the annexing of the humanities by the sciences but a kind of conceptual bridge-building. In fact, I would say that the most consilient—not to mention exciting—work today has come from folks trained as cultural critics—books like Michael Pollan's *The Botany of*

Desire, with its mix of Nietzsche and Richard Dawkins; the sociopolitical sections in Robert Wright's *Non Zero,* and his subsequent writings on the war on terror; Gladwell's work in both *The Tipping Point* and *Blink,* drawing on marketing theory as readily as neuropsychology. (We have also seen the arrival of the consilient blockbuster, in books like Sebastian Junger's *The Perfect Storm,* whose narrative carries the reader all the way from the macro patterns of storm systems in the Atlantic to the molecular interactions that occur in the lungs when humans drown.) My own books have, not surprisingly, explored those same hybrid connections, between the sciences of self-organization and the development of urban culture in *Emergence,* between the neuroscience of social connection and communications theory in *Mind Wide Open.* More cross-disciplinary consilience is no doubt on the way, and it won't come a minute too soon. After two decades of the science wars, we're due for a détente.

Notes

Part One

page xi "Ours is an age besotted with graphic entertainments": George Will, "Reality Television: Oxymoron." http://www.townhall.com/columnists/georgewill/gw20010621.shtml.

page 5 Perhaps most famously, players of Dungeons & Dragons: "Dungeons and Dragons was not a way out of the mainstream, as some parents feared and other kids suspected, but a way back into the realm of story-telling. This was what my friends and I were doing: creating narratives to make sense of feeling socially marginal. We were writing stories, grand in scope, with heroes, villains, and the entire zoology of mythical creatures. Even sports, the arch-nemesis of role-playing games, is a splendid tale of adventure and glory. Though my friends and I were not always athletically inclined, we found agility in the characters we created. We fought, flew through the air, shot arrows out of the park, and scored points by slaying the dragon and disabling the trap. Our influence is now everywhere. My generation of gamers—whose youths were spent holed up in paneled wood basements crafting identities, mythologies, and geographies with a few lead figurines—are the filmmakers, computer programmers, writers, DJs, and musicians of today." Peter Bebergal, "How 'Dungeons' Changed the World," *The Boston Globe,* November 15, 2004.

page 10 Sometimes . . . helpful to imagine culture as a . . . man-made weather system: To be sure, television shows and video games are not water molecules; they come into the world thanks to the passions and talents of individual humans. *Hill Street Blues* needed its Steven Bochco, SimCity its Will Wright. These biographical explanations are not without value, but they are only part of the story. (And of course they are already ubiquitous in the mass media's coverage of themselves, in magazine profiles and newspaper reviews.) But when you're trying to explain macro trends in the history of culture, auteur theory gets you only so far. If Steven Bochco hadn't been around to invent the multithreaded serious drama, someone else would have come along to do it: the economic and technological conditions were too ripe for such an opportunity to be missed.

"Economic and technological conditions" sounds like the neo-Marxist-school cultural materialists, translating each artifact back to the "ultimately determining instance" of material history. But while the cultural materialists did important work in shedding the biographical limits of aesthetic criticism—relating works to their historical moment, and not the vicissitudes of individual genius—they remained too dependent on the symbolic architecture of ideological critique. The work of culture connected to the "economic and technological conditions" the way a mask conveys the face beneath it: representing some common features while distorting others. History churns out a steady progression of new social and technological relations, and culture floats above that world, translating its anxieties and contradictions into a code that, more often than not, makes that experiential turmoil more tolerable to the people living through it. For the kind of criticism at work in this book, on the other hand, the cultural work doesn't attempt to resolve symbolically the contradictions unleashed by historical change. The cultural work is the residue of historical change, not an imagined resolution to it.

page 12 Instead, you hear dire stories: Consider this representative sample of the Trash TV mentality:

"It isn't just nags or fanatics who are disturbed by the harsh new face of TV programming in the late 1990s. Here's what the New York Times had to say in an April 1998 front-page story: 'Like a child acting outrageously naughty to see how far he can push his parents, mainstream television this season is flaunting the most vulgar and explicit sex, language, and behavior that it has ever sent into American homes.' A banner headline in the Wall Street Journal warned not long ago . . . 'It's 8 p.m. Your Kids Are Watching Sex on TV.' U.S. News summarized the trends this way: 'To hell with kids—that must be the motto of the new fall TV season. . . . The family hour is gone. . . . The story of the fall line-up is the rise of sex. Will the networks ever wise up?'

"A wide spectrum of Americans are appalled by what passes for TV entertainment these days. A 1998 poll by the Kaiser Family Foundation found that fully two-thirds of all parents say they are concerned 'a great deal' about what their children are now exposed to on television. Their biggest complaint is sexual content,

followed closely by violence, and then crude language." Karl Zins-
meister, "How Today's Trash Television Harms America," *American Enterprise,* March 1999.

page 12 **"All across the political spectrum":**   Steve Allen, "That's
Entertainment?" *The Wall Street Journal,* November 13, 1998.

page 12 **"The entertainment industry has pushed":**   Parents Television Council. (The passage was found in the past at the Council's website, http://www.parentstv.org/.)

page 12 **"The television sitcom is emblematic":**   Suzanne Fields,
"Janet and a Shameless Culture," *The Washington Times,* February 2, 2004.

page 15 **"The student of media soon comes to expect":**   Marshall
McLuhan, *Understanding Media* (Cambridge, MA: The MIT
Press, 1994), p. 199.

page 17 **"The best that can be said of them":**   Benjamin Spock
and Steven J. Parker, *Dr. Spock's Baby and Child Care* (New York:
Pocket Books, 1998), p. 625.

page 18 **"People who read for pleasure":**   Andrew Solomon,
"The Closing of the American Book," *The New York Times,* July
10, 2004. Solomon is a thoughtful and eloquent writer, but this
essay by him contains a string of bizarre assertions, none of them
supported by facts or common sense. Consider this passage: "My
last book was about depression, and the question I am most frequently asked is why depression is on the rise. I talk about the
loneliness that comes of spending the day with a TV or a computer or video screen. Conversely, literary reading is an entry into

dialogue; a book can be a friend, talking not at you, but to you." Begin with the fact that most video games contain genuine dialogue, where your character must interact with other onscreen characters, in contrast to books, in which the "dialogue" between reader and text is purely metaphorical. When you factor in the reality that most games are played in social contexts—together with friends in shared physical space, or over network connections—you get the sense that Solomon hasn't spent any time with the game form he lambastes. So that by the time he asserts, "Reading is harder than watching television or playing video games," you have to ask: Which video game, exactly, is he talking about? Certainly, reading *Ulysses* is harder than playing *Pac-Man,* but is reading Stephen King harder than playing *Zelda* or *SimCity?* Hardly.

page 24 Invariably these stories point to . . . manual dexterity or visual memory: I don't dwell on the manual dexterity question here, but it's worth noting how the control systems for these games have grown strikingly more complex over the past decade or so. Compare the original *Legend of Zelda* (July 1987), on the original NES, to the current *Zelda,* on the GameCube (March 2003). In sixteen years, games have changed as follows:

THEN	NOW
Controller	*Controller*
4 direction buttons	2 joysticks +
	4 direction buttons
2 action buttons	7 action buttons
Each button has a	Each combo of
single function.	buttons has a
	unique function.

Perspective	*Perspective*
Static overhead view	Dynamic player-controlled "camera" view
You always have complete vision.	Your vision is limited. You must control it.
The game is "flat" (two-dimensional).	The game is "virtual" (three-dimensional).
Gameplay	*Gameplay*
Movement is in one of four directions.	Movement is in any direction, including up and down.
Fighting: 2 buttons	Fighting: More than 10 different button combos. Requires accurate timing and coordination.
Objects: Press a single button.	Objects: Assign a button, learn unique controls to use each object. Requires timing, training.

page 25 So what does the rhinoceros actually look like? Henry Jenkins has painted perhaps the most accurate picture of the rhinoceros of pop culture over the past decade. "Often, our response to popular culture is shaped by a hunger for simple answers and quick actions. It is important to take the time to understand the complexity of contemporary culture. We need to learn how to be safe, critical and creative users of media. We need to evaluate the

information and entertainment we consume. We need to understand the emotional investments we make in media content. And perhaps most importantly, we need to learn not to treat differences in taste as mental pathologies or social problems. We need to think, talk, and listen. When we tell students that popular culture has no place in classroom discussions, we are signaling to them that what they learn in school has little to do with the things that matter to them at home. When we avoid discussing popular culture at the dinner table, we may be suggesting we have no interest in things that are important to our children. When we tell our parents that they wouldn't understand our music or our fashion choices, we are cutting them off from an important part of who we are and what we value. We do not need to share each other's passions. But we do need to respect and understand them." "Encouraging Conversations About Popular Culture and Media Convergence: An Outreach Program for Parents, Students, and Teachers, March–May 2000." http://web.mit.edu/21fms/www/faculty/henry3/resourceguide.html.

page 26 Consider the story of Troy Stolle: Julian Dibbell, "The Unreal-Estate Boom," *Wired*, January 2003.

page 40 Collateral learning in the way of formation: John Dewey, *Experience and Education* (London: Collier, 1963), p. 48.

page 45 "probe, hypothesize, reprobe, rethink":   James Paul Gee, *What Video Games Have to Teach Us About Learning and Literacy.* (New York: Palgrave, 2003), p. 90.

page 63 But another part involves the viewer's "filling in":   There's an old opposition that McLuhan introduced in the early sixties between hot and cool media. I confess that I have long found these categories to be the least useful in the McLuhan canon; there's something counterintuitive about them, something that runs against the grain of the experience they're trying to describe. Hot and cool are defined by the extent to which the audience has to "fill in" the details to complete the information being conveyed. As a medium grows in resolution—and particularly resolution targeted at a specific sense—it requires less participation from the audience, and becomes "hotter" in the process. "A hot medium allows of less participation than a cool one, as a lecture makes for less participation than a seminar, and a book for less than a dialog," McLuhan writes in *Understanding Media* (p. 22). He saw television as a cool medium, partly because of the low resolution of the image itself, and its mosaic style of presenting information. Books, by contrast, were supposed to be hot, and you were left with the unconvincing premise that TV viewers performed more mental labor "filling in" the details than book readers did. Most people, I suspect, would describe it the other way around: books force you to fill in practically everything, because you need to imagine the setting and characters, rather than have them force fed to you through the packaged sound and image on the screen. To me, what's useful in McLuhan's analysis is not hot versus cool, but rather this idea of filling in.

page 65 Multiple threading is the most acclaimed structural convention: For an informative overview of the rise of the multi-threaded drama, see Robert J. Thompson's *Television's Second Golden Age* (Syracuse, NY: Syracuse University Press, 1997).

page 69 The total number of active threads equals the number of multiple plots of *Hill Street*:   The plotlines of *The Sopranos* and *Hill Street Blues* episodes are as follows:

The Sopranos
Christopher's murder
Christopher's screenplay
Conflicts with Uncle Junior
Carmela's frustration
Conflicts with Aunt Livia
Dr. Melfi and Tony
Trouble with the government
Family's finding out what Tony does
Tony's infidelities

Hill Street Blues
Jablonski and the woman
Operation Fleabag
Celestine Gray trial
Renko's paternity
The matricidal iceman
The homicide of the old man
The carjacked tourists
Furillo–Joyce romance

**page 71 The first test screening of the *Hill Street* pilot . . . brought
complaints from the viewers:**   A telling incident occurred at the
end of the show's fifth season, when the production company,
MTM, asked Bochco to leave the series. As an article in *The New
York Times* reported:

"'Hill Street Blues,' the NBC police series that has been acclaimed for its complex narratives and ambitious production techniques, will simplify its plots and reduce the number of characters
next fall in an attempt to lower costs, according to the show's
producers and writers.

"The changes were outlined following the unexpected resignation under pressure last week of Steven Bochco, the show's ground-breaking creator and executive producer. Fewer extras will be used and some regular cast members will appear less frequently than they now do, the show's producers said. They said the changes will help reduce costs and sharpen the image of the series, which in its fifth year reaches 29 percent of the viewers on Thursdays from 10 to 11 p.m.—comfortably above the minimum needed to continue on the network.

" 'The show is probably a little thicker than is good for telling coherent stories,' said Jeffrey Lewis, who along with David Milch was appointed by MTM Enterprises Inc., the producers of the show, to replace Mr. Bochco. 'The problem with the show is we can't tell stories as fully as we like because we have to tell too many.'" Sally Bedell Smith, " 'Hill Street' to Trim Its Cast and Plots," *The New York Times,* March 28, 1985, p. C22.

page 71 First . . . *The Sopranos* is a genuine national hit: With the Season 3 premiere (March 4, 2001), *The Sopranos* began to draw higher audiences than most of its broadcast-network competition, despite its being available in only a third of American households. In particular, it started to routinely smash the competition in the key 18–49 demographic, and frequently still does. For the Season 3 premiere, a 5.8 rating in the 18–49 demographic made it the nineteenth-most-watched program of the week on any network. The Season 4 premiere drew more viewers in its time slot than any other show on television, and episodes during Season 4 routinely beat all broadcast competitors on Sunday nights. For the week overall in the 18–49 demographic, the premiere ranked second, directly behind ABC's *Monday Night Football.*

page 72 Today you can challenge . . . a more complicated mix: In a 1995 interview, Bochco, referring to *Murder One,* clarified his vision for television drama: "What we're trying to do is create a long-term impact. One which requires its viewership to defer gratification for a while, to control that impulse in anticipation of a more complex and fully satisfying closure down the road. It's the same commitment you make when you open up to the first page of a novel." Robert Sullivan, "He Made It Possible," *The New York Times Magazine,* October 22, 1995, p. 54.

page 80 Typical scene from *ER*:   Compare the *ER* dialogue (as appears at http://www.twiztv.com/scripts/attic/er510.htm) with this sequence from a *St. Elsewhere* episode titled "Down's Syndrome." This is the most complicated stretch of medical "texture" in the entire episode, but note how each challenging line is immediately followed by a layperson translation. (The script for this episode, which aired on November 16, 1982, was by Tom Fontana.)

INT. HALLWAY/OUTSIDE MISS TAYLOR'S ROOM—DAY

They stand in the hallway. MORRISON leans against the
wall. WHITE is biting his nails.

WHITE: The liver felt hard, real hard.

AUSCHLANDER: What treatment would you suggest?

ARMSTRONG: Radiation therapy.

AUSCHLANDER: It may relieve some tension but has to
be limited to doses below two thousand rad.

WHITE: How about chemotherapy?

AUSCHLANDER: Again, it might be used in appropriate but futile doses . . . Any other ideas?

MORRISON: What about a partial resection of the liver?

AUSCHLANDER: Some of the best answers don't come from textbooks, Doctor Morrison.

The RESIDENTS look blankly at each other and the floor.

ARMSTRONG: I think she knows she's going to die.

AUSCHLANDER waits for her to continue.

ARMSTRONG: We should try to make her as comfortable as possible. . . . What else can we do?

page 83 But when you watch . . . the other sense of "simpler" applies: "There's a kind of a rule in television," says Jay Anania, a filmmaker who teaches directing at New York University. "You tell people what they're going to see, you show it to them, and then you tell them what they just saw. In *The Sopranos,* nobody clues viewers in to what's about to happen. As in life, there are loose ends that are never tied up. There are metaphors we struggle to divine. [Creator and executive producer David] Chase has said in interviews that he doesn't zoom in on Tony Soprano's face during the protagonist's therapy scenes because he doesn't want to signal to viewers what's important. He wants them to figure that out for themselves." Libby Copeland, "The Sopranos' Four-Octave Range," *The Washington Post,* June 5, 2004.

page 85 Knowing that George uses the alias Art Vandelay: Art Vandelay is referred to in the following episodes: "The Stakeout" (episode 2); "The Red Dot" (episode 29); "The Boyfriend," part 1 (episode 34); "The Pilot," part 1 (episode 63); "The Cadillac," parts 1 and 2 (episodes 124 and 125); "Bizarro Jerry" (episode

137); "Serenity Now" (episode 159); "The Puerto Rican Day" (episode 176); "The Finale," parts 1 and 2 (episodes 179 and 180).

page 86 According to one fan site . . . the average *Simpsons* episode includes: The list of movie references in *The Simpsons* is courtesy the Simpsons Archive website. You can see the entire list at the URL http://www.snpp.com/guides/movie__refs.html. Following is an example of films and their respective references in a "normal" *Simpsons* episode, "Black Widower" (8F20).

The Elephant Man: Lisa's imagination

Cool Hand Luke: picking up garbage; the shot of the chief guard's reflective sunglasses; the guard's cane tapping his leg

The Wizard of Oz: "Snake, I'm going to miss you most of all."

Gone With the Wind: "Fiddle-dee-dee. Tomorrow's another day."

Psycho:

Sideshow Bob turns a chair, expecting to find a corpse, but instead finds Bart. (In the movie, Vera Miles's character turns a chair, expecting to find Mrs. Bates, but instead finds a corpse.)

Sideshow Bob is so startled he hits a swinging lightbulb.

A brief violin sweep shortly thereafter.

The Maltese Falcon: Mary Astor takes the fall (the sliding metal bars of the elevator doors)

Black Widow: Nobody believing the hero's knowledge of the villain; marrying for money, then murdering; the final murder done for revenge; the villain getting overconfident and spilling the beans.

page 92 *Survivor*'s relationship to reality is much closer: *Salon*'s wonderful television critic Heather Havrilesky is one of the few to grasp the fundamental misunderstanding of the "reality" of reality TV: "Many have argued that self-consciousness will be the death of the genre. As more and more contestants who appear on the shows have been exposed to other reality shows, the argument goes, their actions and statements will become less and less 'real.' What's to blame here is the popular use of the word 'reality' to describe a genre that's never been overtly concerned with realism or even with offering an accurate snapshot of the events featured. In fact, the term 'reality TV' may have sprung from 'The Real World,' in which the 'real' was used both in the sense of 'the world awaiting young people after they graduate from school,' and in the sense of 'getting real,' or, more specifically, getting all up in someone's grill for eating the last of your peanut butter." Heather Havrilesky, "Three Cheers for Reality Television," *Salon*, September 13, 2004.

page 94 Some of that challenge comes from . . . the rich social geography: Again, Heather Havrilesky gets it right: "Real people are surprising. The process of getting to know the characters, of discovering the qualities and flaws that define them, and then discussing these discoveries with other viewers creates a simulation of community that most people don't find in their everyday lives. That may be a sad commentary on the way we're living, but it's not the fault of these shows, which unearth a heartfelt desire to make connections with other human beings. Better that we rediscover our interest in other, real people than sink ourselves into the mirage of untouchable celebrity culture or into some überhuman, ultraclever fictional 'Friends' universe." Havrilesky, "Three Cheers for Reality Television."

page 100 "Although the Constitution makes no mention":   Neil Postman, *Amusing Ourselves to Death* (New York: Penguin, 1985), p. 4.

page 119 A decade ago . . . the phrase "screenagers": Douglas
Rushkoff, *Playing the Future* (New York: Riverhead, 1999).

page 121 "Television . . . encompasses all forms": Postman,
Amusing Ourselves to Death, p. 92.

page 121 The second way in which the rise: One way to think
about the cognitive challenge of digital media is through a frame-
work that I outlined in my 1997 book *Interface Culture*. What
makes these new forms uniquely stimulating is that they require
the mastery of interfaces in addition to the traditional "content"
of media, and those interfaces are evolving at a dramatic clip. To
send an e-mail, you need to think about the process of writing, but
also your physical interface with the computer via keyboard and
mouse, the interface conventions that govern the e-mail program
itself, and the larger interface conventions of the operating system.
Compare those different cognitive levels with the more direct system of handwriting a note and you get an idea of the increased
cognitive demands of the modern digital interface.

page 134 On average, Dickens sold around 50,000 copies: Peter
Ackroyd, *Dickens: Public Life and Private Passion* (London: BBC
Worldwide, 2002).

page 136 So this is the landscape of the Sleeper Curve: If pop
music today doesn't appear to be experiencing the same Sleeper
Effect that other mass forms have, that's partly because the repetition revolution already transformed the music industry some
forty years ago, when it switched in the mid-sixties from a business that revolved around throwaway singles to one anchored in
albums designed to be heard hundreds of times. Of course, the

great complexification of popular music that occurred in the sixties had other causes as well—from the talents of individual artists to the volatility of the historical period—but that newfound complexity had room to flower because there was a repetition-friendly format available for artists to explore. Ever since the days of the Victrola, popular music had gravitated to songs that would instantly lodge themselves in listeners' heads, but all that changed in the 1960s. Suddenly the top sellers were long-format albums that rewarded repeated listenings, that offered lyrical and musical complexity unimaginable in the jingle-driven markets that had come before.

In private correspondence, Henry Jenkins points out that a comparable increase in visual and narrative complexity can be seen in the world of comics: "The visual complexity of contemporary mainstream comics would have been nigh on incomprehensible fifty years ago. I say fifty because the push towards visual complexity certainly goes back to the 1960s but an artist today like David Mack or Chris Ware push what a comic page looks like further than would have been imagined by Steranko at his most pop-art inflected wildness. But there is also a new form of narrative complexity which emerges through the development of alternative universes and multiple versions of the same characters. Comics used to develop complexity through continuity—asking readers to keep track of 70 plus years of development in the DC universe, say, and pulling back characters that had not been seen in decades. This is impressive enough—as you suggest in showing similar conduct in contemporary television. But now, they are also allowing different authors to construct radically different versions of the same protagonists, each with their own continuities, each with their own interpretations. So if I am a Spiderman fan, I end up keeping track of four or five different universes each month, recalling as I read an issue whether this is the one where Aunt May

knows about Peter's other identity or not. At the same time, a series like Elseworlds may bend the stories beyond recognition: so
Superman's Metropolis will depict the origins of the Man of Steel
through the language of Fritz Lang's German Expressionist classic or Red Sun will explore what would have happened if the ship
from Kripton landed in the Soviet Union as opposed to the United
States or Speeding Bullets explores what would happen if we
blurred together the origins of Superman and Batman. Each of
these requires extensive knowledge not only of comics but also
[of] a range of other media traditions and the ability to read one
against the other."

Part Two

page 144 If we're not getting these cognitive upgrades: James
Flynn and the economist William Dickens have proposed a fascinating solution for the IQ paradox, one that offers a helpful model
for the gene-culture interaction that has confounded so many
commentators in recent years. "People whose genes send them
into life with a small advantage for these abilities start with a
modest performance advantage. Then genes begin to drive the
powerful engine of reciprocal causation between ability and environment. You begin by being a bit better at school and are encouraged by this, while others who are a bit 'slow' get
discouraged. You study more, which upgrades your cognitive performance, earn praise for your grades, start haunting the library,
get into a top stream. Another child finds that sport is his or her
strong suit, does the minimum, does not read for pleasure, and
gets into a lower stream. Both of you may go to the same school
but the environments you make for yourselves within that school

will be radically different. The modest initial cognitive advantage conferred by genes becomes enormously multiplied.

"Once again, just as different genes are matched with very different environments, so identical genes will be matched with very similar environments. You and your separated identical twin will get very similiar scores on IQ tests at adulthood. Using [Arthur] Jensen's model, genes will get credit for all of the potent environmental influences you both share. And environment will appear so feeble that it could not possibly account for the huge IQ advantage your children enjoy over yourself. Our model shows why this is a mistake. It shows that kinship studies hide or 'mask' the potency of environmental influences on IQ. Therefore, they do not really demonstrate the impossibility of an environmental explanation of massive gains over time." William T. Dickens and James R. Flynn, "Heritability Estimates Versus Large Environmental Effects: The IQ Paradox Resolved," *Psychological Review*, vol. 108, no. 2 (April 2001). A summary can be found at http://www.brookings.edu/views/articles/dickens/200104.htm.

page 146 "The complexity of an individual's environment": Carmi Schooler, "Environmental Complexity and the Flynn Effect," in Ulric Neisser, ed., *The Rising Curve* (Washington, DC: American Psychological Association, 1999), p. 71.

page 152 The *positive* mental impact of contemporary media has not been examined: It's instructive to look at Marie Winn's 1977 book *The Plug-In Drug* in the context of the Flynn Effect. Winn's book—updated in 2002 with additional material critical of the new electronic media—was one of the key original sources of the "television is damaging our children's brains" backlash. In the twenty-fifth-anniversary edition, Winn makes a number of suspect assertions to demonstrate the damaging effects of electronic media. At one point, she admits: "Several generations of children raised watching television have come to maturity showing no signs of a downward trend in overall intelligence" (Marie Winn, *The Plug-In Drug* [New York: Penguin, 2002], p. 67). Technically, of course, this is true. There are no signs of a downward trend because there is, in fact, an *upward* trend. (The Flynn Effect goes unmentioned in the book.)

Winn's primary evidence for the "brain drain" of TV and computers is the long-term trend of declining verbal SAT scores, which she describes as dropping steadily from the mid-sixties to the early eighties, when they then flatline for the next twenty years. She sees this pattern as matching precisely the increasing hourly exposure to television during this period: the generation taking the SAT in 1980 at the very low point of the trend was the first to have been raised on television from cradle to college—and so no wonder that their verbal skills are the worst in recent memory.

Winn's numbers sound convincing, but when you look at them more closely, they strengthen the Sleeper Curve hypothesis more than her brain-drain argument. Where SAT verbals are concerned, the Sleeper Curve prediction would be: A small decline during the heyday of TV, the horrible years of *Happy Days* and *Starsky and Hutch*, followed by a steady but accelerating increase as text-driven interactive media enters the mainstream after 1985 or so.

And, in fact, that's exactly what you see: The *average* verbal SAT score flatlined from 1980 to 2000, but the performance of every single demographic group improved significantly. (Only the overall breakdown of groups changed, lowering the average.) And in the past five years, even the average is up by six points, reflecting the increased emphasis on writing and reading in the digital age.

page 153 One study at the University of Rochester: "Researchers at the University of Rochester found that young adults who regularly played video games full of high-speed car chases and blazing gun battles showed better visual skills than those who did not. For example, they kept better track of objects appearing simultaneously and processed fast-changing visual information more efficiently." Associated Press, "Fire Up That Game Boy," May 28, 2003.

page 153 Another recent study looked at three distinct groups: John Beck and Mitchell Wade, *Got Game?* (Cambridge, MA: Harvard Business School Press, 2004).

page 155 "Just as an elite with a massive IQ": James Flynn, "Massive IQ Gains in 14 Nations: What IQ Tests Really Measure," *Psychological Bulletin*, 101, no. 2 (1987), p. 187.

page 158 In 2003, for the first time, Hollywood made more money: "In 1996, the year before the home DVD player was introduced, consumers spent $6 billion buying VHS tapes, and $9.2 billion renting them, with the studios taking in 75 percent of sales and 20 percent of rentals. In 2004, according to Adams Media Research, consumers will spend $24.5 billion buying and renting DVDs and VHS tapes. Almost $15 billion of that will be in DVD sales, and nearly 80 percent of that will go to the studios through their home entertainment divisions. The explosion in DVD sales has changed the calculus of the Hollywood hit. Last year, 'Finding Nemo' sold $339.7 million in tickets when it was released to the nation's movie theaters. It went on to capture a greater amount—$431 million—in home video (including DVD) retail sales and rentals." Ross Johnson, "Getting a Piece of a DVD Windfall," *The New York Times*, December 14, 2004.

pages 160–161 a philosophy dubbed the theory of "Least Objectionable Programming"; "We exist": Quoted in Thompson, *Television's Second Golden Age*, p. 39.

page 175 "electric speed"; "Today it is the instant speed": McLuhan, *Understanding Media*, p. 353.

page 177 "regime of competence"; "Each level dances": James Paul Gee, "High Score Education," *Wired*, May 2003. The article can be found at http://www.wired.com/wired/archive/11.05/view.html?pg-1.

page 186 "To engage the written word": Postman, *Amusing Ourselves to Death*, p. 51.

page 191 If the subject matter . . . truly had a significant impact on our behavior: The new-media scholar David Gauntlett artfully delineates the problem with the methodology of most media violence studies: "To explain the problem of violence in society, researchers should begin with that social violence and seek to explain it with reference, quite obviously, to those who engage in it: their identity, background, character and so on. The 'media effects' approach, in this sense, comes at the problem backwards, by starting with the media and then trying to lasso connections from there on to social beings, rather than the other way around.

"This is an important distinction. Criminologists, in their professional attempts to explain crime and violence, consistently turn for explanations not to the mass media but to social factors such as poverty, unemployment, housing, and the behaviour of family and peers. The one study that did start at what I would recognise as the correct end—by interviewing 78 teenage offenders (who had been convicted of serious crimes such as burglary and violence) and then tracing their behaviour back towards media usage, in comparison with a group of over 500 'ordinary' school pupils of the same age [Hagell and Newburn, *Persistent Young Offend-*

ers, 1994]—found only that the young offenders watched less television and video than their counterparts, had less access to the technology in the first place, had no particular interest in specifically violent programmes, and either enjoyed the same material as non-offending teenagers or were simply uninterested. This point was demonstrated very clearly when the offenders were asked, 'If you had the chance to be someone who appears on television, who would you choose to be?'

"'The offenders felt particularly uncomfortable with this question and appeared to have difficulty in understanding why one might want to be such a person. . . . In several interviews, the offenders had already stated that they watched little television, could not remember their favourite programmes and, consequently, could not think of anyone to be. In these cases, their obvious failure to identify with any television characters seemed to be part of a general lack of engagement with television' (p. 30)." David Gauntlett, "Ten Things Wrong with the 'Effects Model.'" http:// theory.org.uk/david/effects.htm

page 192 In late 2004, the Departments of Justice and Education released a joint study: Fox Butterfield, "Crime in Schools Fell Sharply over Decade, Survey Shows," *The New York Times,* November 30, 2004.

page 195 "More than any other person, Steven Spielberg":   Kurt Andersen, "Kids Are Us," *The New Yorker,* December 15, 1997.

Notes on Further Reading

page 206 Consilience: In taking a consilient approach to culture, one question invariably arises: Where do you stop? If each step on the ladder connects to another level beneath it, where do you jump off? Why not go from *Zelda*'s problem solving all the way down to quantum gravity? The bestseller lists in recent years have featured a number of books that display precisely that range. (Think of Sebastian Junger's *The Perfect Storm*.) For the critic of popular culture, however, the interpretative ladder has two sensible boundaries, defined by the range of human perception. The scales of reality worth exploring are those that have a material, differential effect on the cultural experience. At the very large and the very small ends of the spectrum, the effects lose relevance. A player may not realize that the video game he's immersed in is activating his dopamine system, but he will feel the effects of that system nonetheless. Some games will generate more dopaminergic activity than others, and as we've seen, games as a genre are more likely to be dopamine-friendly than other cultural forms. So it makes sense to extend our analysis down to the scale of neurochemicals. But the subatomic relationships that ultimately create the dopamine molecule itself are less relevant, because those forces remain constant throughout all brain chemistry, and because their effects are perceived only indirectly.

At the opposite end of the scale, it makes sense to analyze the macroeconomics of the video game industry, because those forces directly shape the kinds of games available to play. But the macro gravitational relationship that allows the earth to revolve around the sun doesn't warrant analysis, because it doesn't have a distinct effect on the game experience. It's true enough that the gaming industry would be dramatically transformed without the sun, but it would be transformed in exactly the same way that all life on earth would be transformed: it would be extinguished. The exact range of appropriate scales varies according to the cultural pursuit in question. If your focus is on the culture of swordfishing, as in Junger's book, then it's entirely appropriate to widen the lens to the global scale of meteorology. But most cultural practices stop at the scale of human collectives: cities, economies, networks. You need to understand how communities now share information online in order to understand the complexity of today's video games. But you don't need to understand the Gulf Stream. As anyone who has tried his hand at this approach will tell you, cutting off the extremes of the ladder hardly limits your perspective. There's plenty of work to do in the middle.

Acknowledgments

This book differs from my previous ones in that its topic is something about which most people have already formed strong opinions. That has its benefits. The many casual conversations one has as one is writing a book turned out, this time around, to be unusually productive. In the past, most of those conversations began with a quizzical look: "You're writing a book about ants and *what?*" But whenever I broached the argument of *Everything Bad*, people would jump into the fray with their own theories about the state of pop culture. Not surprisingly, I found that parents were particularly keen to engage with the ideas. (And sometimes a little suspicious.) Those conversations ended up coloring a great deal of what I eventually wrote: opening up new avenues for exploration, and making me aware of objections that had to be dealt with. So thanks to everyone who chewed me out over a drink or during brunch or on an airplane. You were my imagined readers as I was writing this, for better or worse.

I had a handful of non-imagined readers as well who offered very helpful and supportive comments on the text: Alex Ross, Kurt Andersen, Jeff Jarvis, Henry Jenkins, Douglas Rushkoff, Esther Dyson, Christina Koukkos, Alex Star, and Alexa Robinson. My father managed to find a way to justify all those hours watching *The Sopranos* by making some timely suggestions near the end of the editing process. I am also grateful to Red Burns and George Agudow at NYU's Interactive Telecommunications Program for allowing me to teach a graduate seminar on video games, something no grown adult should rightfully be allowed to do. My students in that seminar were a tremendous help to me in understanding the power and intelligence of the gaming culture.

My editors at *Discover* and *Wired*—Stephen Petranek, Dave Grogan, Chris Anderson, Ted Greenwald, Mark Robinson—let me ruminate on technology and culture in ways that shaped many of the ideas here; Esther Dyson kindly gave me an entire issue of her *Release 1.0* to think about the way software interacts with the brain. I'm grateful as well to the Voices of Vision program at Caltech for inviting me to give a talk on the virtues of pop culture as I was finishing the book.

I'm indebted to my research assistant, Ivan Askwith, who did everything from transcribing book excerpts to generating my (occasionally bizarre) charts to helping me concoct entire theories of *The Sopranos'* narrative universe. I suspect we'll be hearing more from Ivan in the years to come.

What can I say about my editor at Riverhead, Sean Mc-Donald? His new editing technique is unstoppable! I don't think there's a page in this book that wasn't improved by some comment or query of his, and deeply appreciate his willingness to let the book evolve out of the form it took in the original proposal. Thanks to the whole Riverhead team—especially Julie Grau, Cindy Spiegel, Larissa Dooley, Kim Marsar, Liz Connor, and Meredith Phebus—for welcoming me into the fold, and giving me the support and encouragement I needed.

This is the first book I've written from start to finish in our new home in Brooklyn, and so I want to acknowledge the whole supporting cast that makes up the urban oasis that is Park Slope: our many neighborhood friends who dropped by unexpected to save me from a paragraph that couldn't quite find its way to closure; the coffee at Tea Lounge and Naidre's (and yes, Starbucks—everything bad truly is good for you); the hundreds—or thousands—of people who make Prospect Park the perfect spot for an afternoon stroll away from the keyboard; the kids banging away at the study door, demanding some quality time with the computer (and if necessary, with Dad too); and most of all, my wife, who makes so much of the beauty and happiness of our life possible.

But this one is for my agent, Lydia Wills, who has been in the ring with me for ten years now, and who believed in the book when even I had begun to lose faith. If she hadn't become such a superstar over those ten years I might feel as though I owed her something. As it is, I'm just happy she still returns my calls.

New York City
February 2005

ABOUT THE AUTHOR

Steven Johnson is the bestselling author of *Mind Wide Open: Your Brain and the Neuroscience of Everyday Life*; *Emergence: The Connected Lives of Ants, Brains, Cities, and Software*; and *Interface Culture: How New Technology Transforms the Way We Create and Communicate.* He currently writes the "Emerging Technology" column for *Discover* magazine, is a contributing editor to *Wired*, writes for *Slate* and *The New York Times Magazine*, and lectures widely. He lives in New York City with his wife and their two sons. He can be reached on the Web at www.stevenberlinjohnson.com.